T0214821

Lecture Notes in Computer Science   9430

Commenced Publication in 1973
Founding and Former Series Editors:
Gerhard Goos, Juris Hartmanis, and Jan van Leeuwen

Editorial Board

More information about this series at http://www.springer.com/series/8637

Abdelkader Hameurlain · Josef Küng
Roland Wagner (Eds.)

Transactions on Large-Scale Data- and Knowledge- Centered Systems XXII

Springer

Editors-in-Chief

Abdelkader Hameurlain
IRIT, Paul Sabatier University
Toulouse
France

Josef Küng
FAW
University of Linz
Linz
Austria

Roland Wagner
FAW
University of Linz
Linz
Austria

ISSN 0302-9743   ISSN 1611-3349 (electronic)
Lecture Notes in Computer Science
ISBN 978-3-662-48566-8   ISBN 978-3-662-48567-5 (eBook)
DOI 10.1007/978-3-662-48567-5

Library of Congress Control Number: 2015950449

Springer Heidelberg New York Dordrecht London

Printed on acid-free paper

Springer-Verlag GmbH Berlin Heidelberg is part of Springer Science+Business Media
(www.springer.com)

Preface

This volume contains six fully revised selected regular papers. The content of this volume covers a wide range of different and very hot topics in the field of data- and knowledge-management systems. Topics covered include algorithms for large-scale private analysis, modelling of entities from social and digital worlds and their relations, querying virtual security views of XML data, recommendation approaches using diversity-based clustering scores, hypothesis discovery, and data aggregation techniques in sensor network environments.

We would like to express our thanks to the editorial board and the external reviewers for thoroughly refereeing the submitted papers and ensuring the high quality of this volume. Special thanks go to Gabriela Wagner for her high availability and her valuable work in the realization of this TLDKS volume.

July 2015

Abdelkader Hameurlain
Josef Küng
Roland Wagner

Organization

Editorial Board

Contents

BPMiner: Algorithms for Large-Scale Private Analysis

Quach Vinh Thanh and Anwitaman Datta[(✉)]

School of Computer Engineering, Nanyang Technological University,
Singapore, Singapore
{vtquach,anwitaman}@ntu.edu.sg

Abstract. An abundance of data generated from a multitude of sources, and intelligence derived by analyzing the same, has become an important asset across many walks of life. Simultaneously, it raises serious concerns about privacy. Differential privacy has become a popular way to reason about the amount of information about individual entries of a dataset that is divulged upon giving out a perturbed result for a query on a given data-set. However, current differentially-private algorithms are *computationally inefficient*, and do not explicitly exploit the abundance of data, thus wearing out the *privacy budget* irrespective of the volume of data. In this paper, we propose BPMiner, a solution that is both private and accurate, while simultaneously addressing the computation and budget challenges of very big datasets. The main idea is a non-trivial combination between differential privacy, sample-and-aggregation, and a classical statistical methodology called *sequential estimation*. Rigorous proof regarding the privacy and asymptotic accuracy of our solution are provided. Furthermore, experimental results over multiple datasets demonstrate that BPMiner outperforms current private algorithms in terms of computational and budget efficiencies, while achieving comparable accuracy. Overall, BPMiner is a practical solution based on strong theoretical foundations for privacy-preserving analysis on big datasets.

Keywords: Privacy budget · Differential privacy · Sample-and-aggregation · Large-scale analysis

1 Introduction

1.1 Motivation

Recent years have witnessed a tremendous explosion of data generated from different walks of life. The desire to analyze such enormous volumes of data is vast – mining of massive datasets can help us gain valuable insight which leads to significant advances [34]. Consequently, the last few years have witnessed a growing interest both in academia as well as industry to find scalable solutions for large-scale data analytics. The advantages of analytics notwithstanding, associated *privacy* implications is of growing concern. It is desirable to carry out

© Springer-Verlag Berlin Heidelberg 2015
A. Hameurlain et al. (Eds.): TLDKS XXII, LNCS 9430, pp. 1–32, 2015.
DOI: 10.1007/978-3-662-48567-5_1

high-quality analyses, while protecting the confidentiality of information about individuals[1] in a dataset. Privacy protection would help alleviate a data holder's anxiety and also encourage (when individuals may have the choice) him/her to contribute corresponding data for analytics.

Given these issues, a natural question is: *How do we derive intelligence from massive datasets while preserving every individual's privacy?* Differential privacy [14,16] was proposed to address the 'privacy' concern. It prevents compromise of an individual's privacy by ensuring that the presence of his/her data does not significantly change the distribution on the released results, and consequently, the result does not divulge anything more about the individual entries in the dataset. Differential privacy has a competitive advantage over other privacy models: it offers a strong privacy guarantee irrespective of the adversary's background knowledge. Therefore, we employ differential privacy as our desired notion of privacy. Under the umbrella of differential privacy, we also assume the well-known *query-response* model, in which a trusted curator (interactively) responds to multiple queries from the analyst. This model is chosen because (1) it allows derivation of theoretical insight [39] while providing rigor-based privacy assurance, and (2) algorithms under this platform often offer better accuracy (since they focus only on the queries of interest [18]). As a consequence, it has attracted a deluge of works in recent years, e.g., [18,28,33,37,38], and it can also be seen in common frameworks like PINQ [28], SuLQ [8] or GUPT [29].

Computational Efficiency. The 'massive datasets' aspect, on the other hand, is less trivial to address. Literature on differential privacy assumes operations on a statistical database of reasonable size, but the algorithms are not optimized specifically to deal with extremely large datasets. Even without privacy requirements, scaling non-private mining algorithms to big datasets is itself a delicate topic [34]. An advocate for *sampling*, however, may argue that this challenge can be addressed. On handling big data, a simple methodology is *divide-and-conquer*, using sampling to permit computation on relatively small subsets [21,22]. *Sample-and-aggregate (SaG)* [18,29,33,37,38] is a well-known class of differentially private techniques that analyze various samples (blocks) of the original data. As a consequence, SaG algorithms (although not originally intended) seem fitted to work on massive datasets.

However, a deeper investigation invalidates this inference. Current SaG algorithms enjoy strong, rigorous analysis in which the number of blocks increases asymptotically with data size [18,37,38]. Although computation is conducted on a small-sized data block, such a large number of blocks may render the total execution time unacceptable. We illustrate this unfavorable situation with a synthetic dataset of $1,000,000$ rows and 100 columns (which takes up around 800MB, assuming 8 bytes in storage for double floating-point numbers). We generate data in the same way as [22]: for each i-th row, the first 99 columns are

---

[1] In this paper, an 'individual' refers to an entry in a statistical database, which may correspond to information about a real-world entity, e.g., a patient's record, a financial transaction, etc.

$X_i \sim \text{Normal}(0, I)$ where I is a 100×100 identity matrix, and the last column is $Y_i = X_i^T 1 + \epsilon_i$. The computation task is to build a logistic regression model with coefficients constrained in $(-10, 10)$. On a 6-core Intel Xeon E5–1650 running at 3.20GHz with 16GB of RAM, our results show that it takes only 19-20 s to train this model on a single data block. However, the total number of blocks is around 250, so the overall execution time of SaG is approximately 1.3 h. This is undesirable, especially when we consider the fact that 800MB is a relatively small data size. Note that while many of the computation can be parallelized to reduce the absolute time, the net amount of computational resources utilized still stays very high.

Alternatively, one may try to use representative samples of the dataset. In this manner, one can control the data size and make it as suitable as needed for an analytical task at hand. As appealing as this approach sounds, it is by no means straightforward. Impetuous sampling may be unsatisfactory: there is no guarantee that results extracted on a sample is accurate. The question on the sample size requires careful inspection. One may correlate such question with fields like survey sampling, sequential analysis, or sample complexity (learning theory). The first two are often limited to simple statistics like mean, variance, etc., while a (much) larger number of estimators need to be covered. Sample complexity may incur some learning-theoretic measures, e.g. VC or Rademacher complexities, which are sometimes nontrivial to compute. Furthermore, it is static (in that no data is involved during computation), and may result in a bigger sample size than dynamic methods [10].

Budget Efficiency. We now discuss another challenge of private analysis on massive datasets. In practice, an analyst may acquire multiple analyses on data, some of which are unknown beforehand. Unfortunately, the query-response setting does not allow as many analyses as one wants, which stems from the constraint of *privacy budget*[2]. On responding to each query/analysis, the privacy guarantee ϵ is deducted from the budget. Once this budget wears out, further analysis using that data is prohibited; we must discard the database and desist from any future release [19]. In reality, this situation turns out to be a severe issue. Assuming moderate ϵ, posing 5 to 10 queries may exhaust a small privacy budget, and any additional request for analysis will be rejected afterwards. One might wish to reduce ϵ for a larger number of queries; however, smaller ϵ would lead to stricter privacy guarantees, thereby affecting the utility/accuracy of the released outputs.

The issue of privacy budget has existed ever since the inception of differential privacy, but it is amplified with big datasets in the picture. The reason behind this is that current differentially private algorithms assume a rigid model of statistical databases. Specifically, we apply those techniques to all of the data records, *no matter how many* rows there are in the dataset. In other words, whether the given dataset is as small as a few hundreds in size, or whether it

---

[2] While some works regard 'privacy budget' as ϵ (the privacy parameter), we consider it a fixed budget that is reduced per analysis. Such interpretation is seen in [19,28,29].

contains as much as a few billions records, simply makes no difference. Consequently, on a massive dataset, the budget will also run out quickly and further analysis soon rejected. This is somehow a big 'waste' of data, since we have access to a much larger number of records, and yet undesirably restricted to the same amount of analysis.

Accordingly, in this work, we gauge the 'budget efficiency' of an algorithm using the number of analyses/computations it executes until budget exhaustion. As our study of related literature shows, the issues of budget efficiency and computational efficiency remain outstanding challenges for privacy preserved analysis of big data.

1.2 Summary of Contributions

We investigate the intricacies of private analysis of a humongous volume of data, while (1) keeping the process computationally relatively lightweight (i.e., good computational efficiency) and (2) increasing the number of queries that can be answered subject to a given overall privacy budget (i.e., good budget efficiency). We then present a solution framework which we call *BPMiner*. To preserve differential privacy, we use SaG algorithms [18,29,33,37,38]. To resolve the computational challenge, we employ sampling, and determine the 'sufficient' sample size using a classical statistical methodology called *sequential estimation*. Rather than applying it impetuously, we propose a novel modification named *block-moving sequential estimation* to construct estimators with prescribed accuracy. Specifically, blocks are sampled until a pre-defined stopping rule is satisfied, at which point SaG algorithms are applied. By this approach, the computational and accuracy issues are completely addressed. We give theoretical proof of the asymptotic accuracy of our approach, and experimental results show that we have accuracy comparable to SaG while achieving a significant speed-up.

For multiple queries/computations, this process is repeated seamlessly until we run out of data. This may lead to 'intertwining' samples, which is not handled by standard differential privacy, and therefore problematic. We resolve this with (1) a *personal privacy budget* for every data point, and (2) a special privacy definition so-called ϵ-*differential privacy under k-fold adaptive composition*. We provide proof that privacy is still preserved, and empirical evidence that our approach is much more budget efficient, with a high increase in the number of queries.

These ideas will be made clear in subsequent sections. To the best of our knowledge, our algorithm is the first to ever tackle directly the problem of computational efficiency and privacy budget utilization while carrying out private analysis on big data. Another novelty of block-moving sequential estimation is its ability to blend together various ingredients: differential privacy, SaG, and sequential estimation. Our solution works under very mild conditions, and is applicable to the rich class of *(generically) asymptotically normal* estimators [38]. A significance of our solution is that it provides a general (sequential) framework for many SaG algorithms (and possibly any future SaG technique fulfilling those mild conditions). In conclusion, our approach stands out to be a practical solution for any user who wants to utilize a large amount of data for analysis in a privacy-preserving manner.

Related works. To the best of our knowledge, there is no direct related work for large-scale differentially private analysis. There has been a deluge of works in large-scale statistical analysis as well as machine learning on massive datasets [1,12,13,26,41], and many algorithms and architectures have been introduced [23,27,40] (see, e.g., [9] for an empirical evaluation of state-of-the-art frameworks). However, all of them are non-private solutions that are not fitted to address the problem at hand. Even if one somehow managed to extend those solutions so that they could provide privacy protection, the fact that they use all of the available data would wear out the privacy budget quickly (as mentioned before). Meanwhile, there are also many algorithmic solutions from the privacy perspective, i.e., algorithms that help preserve differential privacy for participating users [18,29,37,38] (see [15] for a comprehensive survey). These solutions, however, are not proposed to work in the setting of large datasets; specifically, they cannot address the issues of computational and budget efficiency mentioned.

Significance and applications. Our solution provides a bridge between large-scale analysis and differential privacy. It helps analysts to conduct data analysis on large datasets while protecting the confidentiality of information about participating users. One significance of our solution is that we do not use the whole dataset per analysis. Besides being more computationally efficient, it can help analysts to obtain *approximate* (similar to [2,42,43]) or *early* results (similar to [24,25]) that are good approximations to the true results. This is particularly desirable when analysts face a huge amount of data but want to apply some analysis and view results quickly. In addition, because of analyzing only a portion of available data, our solution might also be particularly useful in settings where computation is priced and smaller amounts of computation are favored. This is common in the era of cloud computing where computation is provided as a service. Finally, we want to emphasize that on top of such advantages is a layer of privacy protection, which essentially helps data owners feel much more secure about contributing their data to analysis.

1.3 Outline

We note that there is no specific work which addresses precisely the same set of issues as we do, but relevant related works are already mentioned while motivating the scope of this work, and also in Sect. 2 where we delve into the details of some preliminaries and notation on which the current work builds upon.

The rest of the paper is organized as follows. We formally define in Sect. 3 what we mean by 'privacy' and 'accuracy'. Section 4 then introduces the novel principles of BPMiner, while Sect. 5 presents the details of our main algorithms which follow nicely from such principles. In Sect. 6, we provide theoretical analysis of why our solution is both private and accurate, while extensive experimental results on accuracy, computational efficiency and budget efficiency of the solution are also reported. Finally, Sect. 7 concludes with several future research directions.

2   Preliminaries and Notation

Consider a domain $\mathcal{D}$ of data points, and let $\mathcal{D}^*$ contain all subsets of $\mathcal{D}$. We model the dataset/database by a vector-valued random variable $X \in \mathcal{D}^*$. Typically, $X = X_1, \ldots, X_n$, where $X_i \in \mathcal{D}$ are independent and identically distributed (IID) according to some distribution P. Suppose we need to estimate some parameter of P, say θ. This 'truth value' θ is called the *estimand* and unknown to us. We assume θ takes values in the parameter space $\Theta = \mathbb{R}^d$.

An *estimator/statistic* $T : \mathcal{D}^* \to \Theta$ is a function $T(X)$ defined over $\mathcal{D}^*$, also taking values in Θ. An estimator T is constructed based on X and therefore random. When a realization $X = x$ of the data is observed, the value $T(x)$ is called an *estimate*. If we could have a good estimator $T(X)$, then $T(x)$ would be our educated guess for θ.

In some cases, we may need to construct an estimator $\hat{\theta}$ on the basis of another estimator T. For example, [29,38] used asymptotically normal statistics to create a new private estimator. Under such circumstances, we use the notation $\hat{\theta}_T$ to emphasize that $\hat{\theta}$ essentially depends on T for its construction. In this paper, $\hat{\theta}_T(X)$ represents the final estimator that we seek out.

2.1   Preliminaries from Statistics

Convergence in Probability and CLT. A sequence of random variables $Y_n, n = 1, 2, \ldots$ converges to some constant c *in probability*, denoted $Y_n \overset{P}{\to} c$, if for every $\epsilon > 0$, $\Pr\left(|Y_n - c| < \epsilon\right) \to 1$ as $n \to \infty$. In addition, we will also use the well-known *Central Limit Theorem (CLT)*: if $Y_1, \ldots, Y_n$ be i.i.d. random variables with expectation $\mathbb{E}Y_i = \mu$ and variance $\text{Var}(Y_i) = \sigma$, then as $n \to \infty$, $(\sqrt{n}(\bar{Y} - \mu))/\sigma \overset{D}{\to} N(0, 1)$, where $\overset{D}{\to}$ denotes convergence *in distribution*, and $\bar{Y} = n^{-1} \sum_{i=1}^{n} Y_i$ is the sample mean.

Kolmogorov-Smirnov (KS) Distance. To measure the closeness between two probability measures P and Q on $\mathbb{R}^d$, we use the KS distance, defined by $\text{KS}(P, Q) = \sup_R |P(R) - Q(R)|$, where R are axis-parallel rectangles [38]. It is generally known that the KS distance is stronger than convergence in distribution. Specifically, if $\text{KS}(P_n, P)$ tends to 0 as $n \to \infty$, then P_n converges in distribution to P. Like [38], we use extended notation for the distance between random variables. For example, $\text{KS}(X_n, X)$ is the KS distance between the random variables X_n and X.

Generic Asymptotic Normality. Generic asymptotic normality is introduced in [38] to indicate estimators having three properties: asymptotic normality, linear bias, and bounded third moment. For some value σ_P which depends on

P, if a statistic T satisfies (1) $\sqrt{n}(T(X) - \theta)/\sigma_P \xrightarrow{D} N(0,1)$ as $n \to \infty$, (2) $\mathbb{E}[T(X)] - \theta = O(1/n)$, and (3) $\mathbb{E}\left(\sqrt{n}|T(X) - \theta|/\sigma_P\right)^3 = O(1)$, then T is said to be *generically asymptotically normal* at distribution P [38].

2.2 Differential Privacy

ϵ-Differential privacy (ϵ-DP) [14,16] is proposed to protect information about individuals in a statistical database.

Definition 1 (ϵ-DP). *A (randomized) algorithm A gives ϵ-differential privacy if for all neighboring datasets X and X' (i.e., differing on at most one element), and all subsets $S \subset \mathsf{Range}(A)$,*

$$\Pr(A(X) \in S) \leq \Pr(A(X') \in S) \times e^\epsilon,$$

where $\epsilon > 0$ the privacy parameter.

A differentially private output's distribution is essentially the same regardless of an individual's presence in its calculation. In other words, individuals have low impact on the distribution of released results, disorienting the analyst/adversary. There has been a plethora of works on how to construct differentially private algorithms. Results on differential privacy may be found in [15].

2.3 Sample-and-Aggregate

Sample-and-Aggregate (SaG) is a methodology which preserves differential privacy. It was proposed in [33] and further instantiated by many follow-up algorithms [18,29,37,38]. Procedure 1 presents the basic framework for SaG. Each algorithm proposed a different Range. [18,37] used the diameter of Θ. [38] used private quantiles to construct an approximated interquartile range. [29] suggested 'GUPT-tight', 'GUPT-loose' and 'GUPT-helper'.

Algorithm 1. The Sample-and-Aggregate (SaG) framework

Input: Dataset $X = (X_1, \ldots, X_n)$, Privacy parameter ϵ, Computation T
Output: An ϵ-differentially private estimator $A_T(X)$

1 Randomly divide n data points of X into k disjoint blocks $B_1, \ldots, B_k$.
2 Compute $Z_i = T(B_i)$ for each $i = 1, \ldots, k$.
3 Compute the average: $\bar{Z}_k = k^{-1} \sum_{i=1}^{k} Z_i$.
4 Compute Range and sample $Y \sim \mathrm{Laplace}\left(\frac{\mathsf{Range}}{k\epsilon}\right)$.
5 Output the estimator $A_T = \bar{Z}_k + Y$.

Although $\bar{Z}$ is constructed based on the statistic T, we make an exception and do not use the notation $\bar{Z}_T$. Instead, the subscript T is implicitly understood, and we simply write $\bar{Z}$. In addition, whenever the number of blocks k is important for our discussion, we will use $\bar{Z}_k$ (as in Algorithm 1).

2.4   Sequential Estimation

Sequential estimation is applied in scenarios where we want better data usage. Unlike classic estimation, the sample size that we use to construct estimators is not known beforehand. Our purpose is to find the *optimal* sample size, denoted by n^*, that is the minimum number of data points for which estimation is good. n^* can be defined by rules of the form (see [3,11,36]):

$$n^* = \text{Smallest integer } n \geq [\text{Expression}]. \tag{1}$$

Usually, Expression involves unknown parameters, and we approximate them with either multi-stage or purely-sequential methods. In *multi-stage* methods, we sample some *pilot data* during the first stage, and then use it to approximate the parameters; an example is Stein's two-stage procedure [31,36]. In *purely-sequential* methods, we keep adding data points (while approximating the parameters on what we've sampled) until the condition is satisfied; examples include [3,11,31].

Regardless of what method we use, the parameters are always approximated based on the data we have in hand. Consequently, the resulting approximation of n^* is a random variable, and we use the capital letter N to denote it. The (approximated) rule is:

$$N = \text{Smallest integer } n \geq [\text{ApproximatedExpression}]. \tag{2}$$

An interested reader may seek [31] for more details.

3   Privacy and Accuracy

A critical requirement for an acceptable solution is high utility/accuracy. In this section we provide formal definitions of what we mean by a 'private' and 'accurate' solution.

3.1   Privacy

For a **single computation**, we require the algorithm to be ϵ-differentially private (as defined in Definition 1). Unfortunately, complication arises when we need to perform **multiple computations**. Composition theorems [28] specify that privacy guarantees are degraded in a well-controlled manner. However, such argument deals with multiple mechanisms on the *same* database.

In this paper, we use sequential estimation to make better use of the privacy budget. This essentially involves sampling and executing (differentially private) mechanisms on the sample. When faced with multiple computations, we might choose a different sample per computation. This situation would lead to intertwining samples, and be problematic due to two reasons. First, a data point may be involved in several computations. In this case, we are uncertain about how the budget is affected, since a 'privacy budget' is allegedly applicable to the whole database. Second, the situation is not covered by standard composition theorems; therefore, we have no idea how privacy is degraded.

To deal with this clutter, we use ϵ-*differential privacy under k-fold adaptive composition (kFold-ϵ-DP)*, a variant of differential privacy in [17]. This privacy guarantee concentrates on how an individual's data is exposed to multiple analyses over its lifetime.

Definition 2 (kFold-ϵ-DP). *Let $\mathcal{A}$ be a family of (randomized) algorithms. Let k-fold composition experiment b (b $\in \{0,1\}$) be that: for any $i = 1,\ldots,k$, an adversary Adv outputs two neighboring datasets X_i^0, X_i^1, together with some algorithm $A_i \in \mathcal{A}$, and receives $y_i = A_i(X_i^b)$. The family $\mathcal{A}$ is said to be ϵ-differentially private under k-fold adaptive composition if for every adversary Adv, we have*

$$D_\infty(\mathit{View}^0 \| \mathit{View}^1) \leq \epsilon,$$

where View^0 and View^1 are the views of Adv in the k-fold composition experiment 0 and 1, respectively. We define $D_\infty(Y\|Z) = \max \ln(\Pr(Y = y)/\Pr(Z = z))$.

We refer the reader to [17] for more details. Informally, if a family of mechanisms has ϵ-DP-kFold, the adversary cannot tell whether an individual's data is used during the k experiments. ϵ-DP-kFold gives a reasonable analogue of differential privacy when individuals are engaged in multiple databases and mechanisms. For this reason, we postulate that a solution should preserve ϵ-DP-kFold.

3.2   Accuracy

Our estimator $\hat{\theta}(X)$ is random. Therefore, when saying $\hat{\theta}(X)$ is 'accurate', we expect it to *close* to θ *with high probabiblity*. By 'with high probability', we postulate that the event in consideration should happen with probability not less than $1 - \alpha$, where α is a small positive number. By 'close', we consider two interpretations. In the *absolute* interpretation, $\hat{\theta}_T(X)$ is close θ if its absolute deviation from θ, $|\hat{\theta}_T(X) - \theta|$, is small (i.e., not greater than some small threshold δ). In other words, $\hat{\theta}_T(X)$ is said to be *absolute (δ, α)-accurate* if

$$\Pr(|\hat{\theta}_T(X) - \theta| \leq \delta) \geq 1 - \alpha. \tag{3}$$

In the *relative* interpretation, $\hat{\theta}_T(X)$ is close θ if the relative deviation $(|\hat{\theta}_T(X) - \theta|)/|\theta|$ is less than δ. That is, if

$$\Pr(|\hat{\theta}_T(X) - \theta| \leq \delta|\theta|) \geq 1 - \alpha, \tag{4}$$

then $\hat{\theta}_T(X)$ is called *relative (δ, α)-accurate*. Both interpretations are intuitive and common. They are considered standard in sequential estimation [10,11,31,32], and closely related to confidence intervals or PAC learning. However, we deem those definitions impractical.

For absolute (δ, α)-accuracy, the absolute error may become unnecessarily small when θ is large. For example, suppose $\theta = 10^5$ is the population mean that we aim to estimate. Demanding that $|\hat{\theta}_T(X) - \theta| \leq \delta = 0.1$, or $\hat{\theta}_T(X) \in [9999.9, 10000.1]$, is simply overkill. A suitable choice of δ requires certain knowledge about the domain (specifically, θ). Meanwhile, relative (δ, α)-accuracy is less domain-dependent, alleviating the above problem. However, another issue arises when θ is close 0. For example, say $\theta = 0.05$, then a small δ makes $\delta|\theta|$ even smaller. In order to satisfy the condition (4), we would need a substantially large sample. When θ is exactly 0 (e.g., mean of a standard normal population), the definition becomes invalid.

We resolve the above issues by requiring our estimator, $\hat{\theta}_T(X)$, to satisfy a mixed definition, so-called (d, α)-accuracy.

Definition 3 ((d, α)-accuracy). *An estimator $\hat{\theta}_T(X)$ is said to be (δ, α)-accurate if*

$$\Pr(|\hat{\theta}_T(X) - \theta| \leq \max(\delta, \delta|\theta|)) \geq 1 - \alpha,$$

for some small number $\alpha \in (0, 1)$ and $\delta > 0$. Here α is the confidence parameter, and δ is the accuracy parameter. The probability is taken w.r.t. both randomness in $\hat{\theta}_T$ (i.e., its coin flips) and randomness in sampling of X.

We highlight that (δ, α)-accuracy addresses the limitations of both 'absolute' and 'relative' definitions. For example, consider $\delta = 0.05$. When $\theta = 10^5$, we need $|\hat{\theta}_T(X) - \theta| \leq \delta|\theta|$, or $\hat{\theta}_T(X) \in [9500, 10500]$. When $\theta = 0.8$, we require that $|\hat{\theta}_T(X) - \theta| \leq \delta$, or $\hat{\theta}_T(X) \in [0.75, 0.85]$. We believe that these intervals are appropriate; for instance, estimates of 9750 and 0.785 are reasonably good relative to $\theta = 10^5$ and $\theta = 0.8$, respectively.

3.3   Goal

The goal of this paper is to construct an algorithm that satisfies these definitions of privacy and accuracy in the massive-data setting, while simultaneously addressing the challenges of computational and budget efficiency. In the next two sections, we propose such a solution, BPMiner.

4   Principles of BPMiner

4.1   A Simple Approach

SaG algorithms operates by partitioning the dataset X into multiple blocks $B_1, \ldots, B_k$. Consequently, they use all data per computation, which would quickly exhaust the privacy budget. In addition, they are not scalable, as illustrated in Sect. 1. A remedy for these issues is sequential estimation.

Sequential estimation w.r.t. data points. The idea is simple: To make efficient use of data, we first use sequential estimation to find the (approximated) optimal sample size N, as mentioned in Sect. 2.4. Once N is obtained, we apply SaG methods on a sample of size N. In this manner, data are not used up per computation, which increases computational efficiency and 'fattens' the privacy budget.

This strategy is a straightforward adoption of sequential estimation, in which N is essentially the optimal number of *data points*. For this reason, we call the approach *point-moving sequential estimation*. We highlight that such naïve adoption has two limitations. First, sequential estimation does not support arbitrary statistics: it is mostly concerned with estimating the population mean. Meanwhile, we want to work with a much larger class of estimators.

Second, sequential estimation ensures that estimating on a N-sized sample, say $(X_1, \ldots, X_N)$, is good. However, the estimator should be applied 'directly', that is, used over the N-sized sample (e.g., $\hat{X}_N = N^{-1} \sum_{i=1}^{N} X_i$ for the mean estimator). We, however, have little (theoretical) guarantee that applying sample-and-aggregate on this N-sized sample would also lead to good results.

4.2   Assumptions

Before delving into the details of BPMiner, we first state the set of conditions under which it works. We will soon see that these assumptions are necessary to address the above limitations.

First, the statistic $T(X)$ needs to have **linear bias**, i.e., $\mathbb{E}[T(X)] - \theta = O(1/n)$. Essentially, $T(X)$ must have low bias for large n. In reality, we often favor small bias, so this assumption is easily fulfilled by many practical estimators. An important example is the (rich) class of generically asymptotically normal statistics [38], as mentioned in Sect. 2.

The second condition is that the 'aggregate' phase in sample-and-aggregate is conducted by (simple) **averaging**. This assumption turns out to be quite common: Algorithm 1 illustrates how averaging has been used in most SaG methods [18, 29, 37, 38].

It can be highlighted that BPMiner works under very mild and practical assumptions, which makes it highly applicable.

4.3   Key Strategy of BPMiner

The idea behind our strategy is to apply sequential estimation in a non-standard manner. Let us first consider the set $Z = (Z_1, \ldots, Z_k)$ produced in the SaG framework (see Algorithm 1). Since $Z_i, i = 1, \ldots, n$ are i.i.d., we can regard Z as a dataset induced on X. Let $\mathbb{E}Z$ ($= \mathbb{E}Z_1 = \cdots = \mathbb{E}Z_k$) be the population mean, and $\bar{Z} = k^{-1} \sum_{i=1}^{k} Z_i$ the sample mean.

We previously mentioned that the mean is commonly used in sequential estimation, so $\mathbb{E}Z$ would be a good fit. Furthermore, the averaging assumption allows us to work with $\bar{Z}$. Therefore, a suitable choice is to aim attention at $\mathbb{E}Z$ and $\bar{Z}$: first, we use sequential estimation on Z to find the optimal size, and then we compute $\bar{Z}$ as an estimate of $\mathbb{E}Z$. This is the idea of *sequential estimation w.r.t. data blocks*, presented in Sect. 4.3.

The original requirement, however, is to ensure the closeness between θ and $\hat{\theta}_T$. While θ and $\mathbb{E}Z$ are nicely (asymptotically) close by the linear bias assumption, dealing with $\hat{\theta}_T$ is not as straightforward. SaG methods construct $\hat{\theta}_T$ by perturbing $\bar{Z}$ with Laplace noise, so we need to 'shift' accuracy requirements of $\hat{\theta}_T$ to $\bar{Z}$. This idea about *shift of prescribed accuracy* is discussed in Sect. 4.3.

We also argue that this strategy is highly favorable, and we shall discuss a few of its competitive advantages in Sect. 4.3.

Sequential Estimation W.r.t. Data Blocks. The idea is to work in terms of *data blocks*. The number of blocks k in Algorithm 1 is no longer fixed in advance. Instead, we keep sampling data blocks and producing block estimates, while treating these estimates as 'data points' in the context of sequential estimation. In this manner, the optimal sample size is the minimum number of blocks for which we obtain good results. We use k^* to denote this value.

$$k^* = \text{Smallest integer } k \geq [\text{Expression}]. \tag{5}$$

Similar to Sect. 2.4, we approximate the unknown parameters in Expression using either multi-stage or purely-sequential methods. The resulting approximation of k^* is obtained based on data, and therefore a random variable. We denote it by the capital letter K:

$$K = \text{Smallest integer } k \geq [\text{ApproximatedExpression}]. \tag{6}$$

It can be seen that (5) and (6) are very similar to (1) and (2), respectively. The difference is that the (approximated) optimal sample size K (or k^*) is now associated with data blocks rather than data points. For this reason, we call this idea *block-moving sequential estimation*[3].

Shift of Presribed Accuracy. SaG algorithms aim to construct the estimator $\hat{\theta}_T = \bar{Z} + Y$, where the Laplace noise Y is chosen such that $\text{Var}(Y)$ is small [18, 37, 38]. In this manner, Y insignificantly affects the distribution of $\bar{Z}$, thereby explaining the good performance of $\hat{\theta}_T$ in those works. As a consequence, we expect similar asymptotic behaviour from $\bar{Z}$ and $\hat{\theta}_T$, as formally stated in the following theorem.

Theorem 1. *Let $\hat{\theta}_T(X)$ denote an estimator from* [18, 29, 37, 38]. *Then as $k \to \infty$, $\text{Pr}(|\hat{\theta}_T(X) - \theta| \leq \delta) \to \text{Pr}(|\bar{Z} - \theta| \leq \delta)$.*

The proof is provided in the Appendix. By Theorem 1, we can shift the prescribed accuracy from $\hat{\theta}$ to $\bar{Z}$ for previous sample-and-aggregate methods. This theorem allows us to concentrate our efforts on $\bar{Z}$ rather than $\hat{\theta}_T$.

Advantages. In comparison to point-moving sequential estimation, the proposed strategy is superior in three respects. First, the above discussion implies

---

[3] This name reflects the fact that we use sequential estimation w.r.t. data blocks. It is not to be confused with moving blocks boostrap in the field of bootstrap/subsampling.

that working in terms of blocks conforms greatly to sequential estimation. By focusing on $\mathbb{E}Z$ and $\bar{Z}$, we can enjoy well-established technicalities of estimating the mean in sequential estimation. This guarantees the accuracy of $\bar{Z}$, and also of $\hat{\theta}_T$ due to the shift in Sect. 4.3. Overall, working w.r.t. data blocks helps us address the two issues of point-moving sequential estimation, specified in Sect. 4.1.

Second, block-moving sequential estimation also conforms to SaG. Once K is found, the list of built estimates $(Z_1, \ldots, Z_K)$ is also ready for use. At this point, the 'sample' stage is already finished; we simply 'aggregate' $(Z_1, \ldots, Z_K)$ to obtain the final estimate $\hat{\theta}_T$. Compared to point-moving sequential estimation, where N is found *first* and SaG algorithms are employed *afterwards*, we require much less computation and data scan. This contributes greatly to the (computational) efficiency of our solution.

Finally, our strategy blends together analysis of both sequential estimation and SaG. The number of blocks k is prominent when analyzing the asymptotic behavior of SaG algorithms [18,37,38]. Meanwhile, sequential estimation considers asymptotic evaluations when the accuracy parameter d tends to 0, or equivalently, when K increases without bound. We, therefore, can kill two birds with one stone when analyzing a large number of blocks.

4.4 Personal Privacy Budget

As discussed, we employ the idea of sequential estimation to sample enough data blocks for estimation. This, however, is w.r.t. a single computation. With multiple computations, an individual's data is involved in different and intertwining samples. In this case, we require ϵ-differential privacy under k-fold adaptive composition, as presented in Sect. 3. To comply with this notion of privacy, we focus on an individual's perspective, and postulate the concept of *personal privacy budget*. We can interpret each individual's personal budget as his/her own tolerance of privacy. The concept is arguably reasonable, since some individuals may be more relaxed about their confidentiality than others.

Suppose a computation requires some privacy guarantee ϵ. On handling this computation, not all budgets are affected. Instead, we deduct ϵ from an individual's budget only when his/her record contributes to the analysis. Also, when an individual's budget wears out, that record no longer participates in any future computation.

We highlight a significance of introducing personal privacy budget. When new records come in, we append them to the present data, with their budgets 'untouched'. If some records are removed from the database, it is hardly problematic since we simply use blocks sampled from the rest of the data. These observations show that our approach is highly applicable to *dynamic* databases (e.g., stream of data).

5 Algorithms

We recall the strategy: use sequential estimation to find the optimal number of blocks k^*, and apply the SaG framework as a finishing touch. Since k^* involves unknown parameters, we approximate it by some number K obtained based on data. We shall derive the formula for k^* in Sect. 5.1. We provide two approximations for k^*: K_{Seq} in Sect. 5.2 (using the purely-sequential technique) and K_{Mul} in Sect. 5.3 (using the multi-stage technique). The previous notation for approximations of k^*, K, serves as a general term for both K_{Seq} and K_{Mul}. Section 5.4 introduces an algorithm to manage the budgets of all data points. Finally, all elements of the solution are put together in Sect. 5.5.

As mentioned in Sect. 4.3, once K is found, the estimates $(Z_1, \ldots, Z_K)$ is also readily available: the 'sample' part of sample-and-aggregate has been accomplished. For this reason, we name the algorithms that find K_{Seq} and K_{Mul} Seq-*Sampler* and Mul-*Sampler*, respectively.

5.1 Derivation of k^*

Absolute (δ, α)-accuracy. We first consider absolute (δ, α)-accuracy, where our goal is to compute the optimal size k^* such that (3) is satisfied. Let us examine some number of blocks k. The list of produced estimates $(Z_1, \ldots, Z_k)$ is a dataset with i.i.d. Z_i, $i = 1, \ldots, k$. Let μ and σ^2 denote the mean and variance of Z_i, respectively. The shift of prescribed accuracy in Sect. 4.3 allows us to approximate $\Pr(|\hat{\theta}_T(X) - \theta| \le \delta)$ by $\Pr(|\bar{Z} - \theta| \le \delta)$. Combining this with the accuracy requirement (3), it turns out that we need to guarantee

$$\Pr\left(|\bar{Z} - \theta| \le \delta\right) \approx 1 - \alpha. \tag{7}$$

Recall the basic equality $|a - b| = ||a - c| \pm |b - c||$. The left-hand side of (7) is therefore

$$\Pr\left(\left|\sqrt{k}\frac{|\bar{Z} - \mathbb{E}\bar{Z}|}{\sigma} \pm \sqrt{k}\frac{|\theta - \mathbb{E}\bar{Z}|}{\sigma}\right| \le \frac{\delta\sqrt{k}}{\sigma}\right).$$

Let us examine the second term $\sqrt{k}|\theta - \mathbb{E}\bar{Z}|/\sigma$. Since $Z_1, \ldots, Z_k$ are i.i.d., $\mathbb{E}\bar{Z} = k^{-1}\sum_{i=1}^{k}\mathbb{E}Z_i = \mu$. By the linear bias assumption, this is $\theta + O(1/b)$, where b is the common block size for which we build $Z_1, \ldots, Z_k$. The numerator $|\theta - \mathbb{E}\bar{Z}|$ is thus $O(1/b)$.

We can make the second term vanish asymptotically by choosing b such that $1/\delta = o(b)$ (see Sect. 6.1). Supposing b is chosen like this, we are left with the first term, i.e.,

$$\Pr\left(\sqrt{k}|\bar{Z} - \mu|/\sigma \le \delta\sqrt{k}/\sigma\right) \approx 1 - \alpha. \tag{8}$$

The CLT gives us that $\sqrt{k}(\bar{Z} - \mu)/\sigma$ is approximately a standard normal random variable: $\sqrt{k}(\bar{Z} - \mu)/\sigma \xrightarrow{D} N(0, 1)$ as $k \to \infty$. Let $\Phi(\cdot)$ be the cumulative distribution function (CDF) of a standard normal random variable, then the left-hand side of (8) is approximately $2\Phi(\delta\sqrt{k}/\sigma) - 1$. Equating this with $1 - \alpha$ gives $\Phi(\delta\sqrt{k}/\sigma) = 1 - \alpha/2$. Let $z_{\alpha/2} = \Phi^{-1}(1 - \alpha/2)$. Solving for k gives us $k = \sigma^2 (z_{\alpha/2}/\delta)^2$. This is the optimal value of k required to guarantee the accuracy of $\bar{Z}$ (or equivalently, $\hat{\theta}_T$). In other words, this is exactly k^* that we want.

Lemma 1. *For absolute* (δ, α)-*accuracy, the required optimal size is:* $k_a^* = small$-*est integer* $k \geq C_1 = \sigma^2 (z_{\alpha/2}/\delta)^2$.

Relative (δ, α)-**accuracy.** We aim to derive the formula of k^* for relative (δ, α)-accuracy. Akin to (7), we need

$$\Pr\left(|\bar{Z} - \theta| \leq \delta|\theta|\right) \approx 1 - \alpha. \tag{9}$$

Similar to above, the left-hand side of (9) is

$$\Pr\left(\left|\sqrt{k}\frac{|\bar{Z} - \mathbb{E}\bar{Z}|}{\sigma} \pm \sqrt{k}\frac{|\theta - \mathbb{E}\bar{Z}|}{\sigma}\right| \leq \left|\frac{\delta\sqrt{k}}{\sigma}\theta\right|\right). \tag{10}$$

Again, the second term vanishes when $1/\delta = o(b)$. We now handle the third term. The linear bias assumption gives us $\mu = \theta + O(1/b)$, which translates to $\theta = \mu + O(1/b)$ due to the definition of $O(\cdot)$ order. Therefore, the third term of (10) is $\left|\frac{\delta\sqrt{k}}{\sigma}\mu + \frac{\delta\sqrt{k}}{\sigma}O(1/b)\right|$. Meanwhile, $1/\delta = o(b)$ makes sure that $O(1/b) \times \delta\sqrt{k}/\sigma$ tends to 0 (see Sect. 6.1). We are thus left with $|\mu\delta\sqrt{k}/\sigma|$. Combining all arguments, (9) becomes $\Pr(|\bar{Z} - \mu|\sqrt{k}/\sigma \leq |\mu|\delta\sqrt{k}/\sigma) \approx 1 - \alpha$. We proceed the derivation in a similar way to that of absolute (δ, α)-accuracy, with δ now being replaced by $\delta|\mu|$. Note that the asymptotic evaluations remain unaffected since $|\mu|$ is fixed. We have the following lemma.

Lemma 2. *For relative* (δ, α)-*accuracy, the required optimal size is:* $k_r^* = small$-*est integer* $k \geq C_2 = \sigma^2 (z_{\alpha/2}/\delta|\mu|)^2$.

(δ, α)-**accuracy.** We now discuss the main accuracy requirement in Definition 3, which blends together absolute and relative (δ, α)-accuracy. The smaller δ is, the larger k_a^* and k_r^* need to be so that prescribed accuracy is satisfied. In (δ, α)-accuracy, we require the accuracy parameter $\max(\delta, \delta|\theta|)$, i.e., the 'lighter' between δ and $\delta|\theta|$. As a consequence, we only need the 'lighter' between the two optimal sizes k_a^* and k_r^* such that $|\hat{\theta}_T(X) - \theta| \leq \max(\delta, \delta|\theta|)$. In other words, $k^* = \min(k_a^*, k_r^*)$. Note that this argument can also be seen in [32].

Theorem 2 (Optimal size k^*). *The required optimal size k^* for which $\hat{\theta}$ is* (d, α)-*accurate is:* $k^* = \min(k_a^*, k_r^*) = smallest$ *integer* $k \geq \min(C_1, C_2)$.

Approximation of k^*. In order to find k^*, we need to calculate k_a^* and k_r^*, or equivalently, C_1 and C_2. Unfortunately, these values cannot be computed since they involve unknown parameters: σ^2 in C_1, σ^2 and μ in C_2. Therefore, we need to approximate C_1 and C_2, which in turn translates to the problem of approximating σ^2 and μ.

Consider a list of estimates $(Z_1, \ldots, Z_k)$. In general, we approximate μ with the sample mean $\bar{Z}_k$, and σ^2 with the *sample variance*: $S_k^2 = \frac{1}{k-1} \sum_{i=1}^{k} (Z_i - \bar{Z})^2$. Computation of $\bar{Z}_k$ and S_k^2 requires knowledge of k. Unfortunately, k is not known beforehand in sequential estimation, making such approximations not readily available. Below we present two methods to resolve this issue: purely-sequential (Seq-Sampler) and multi-stage (Mul-Sampler).

The strategy to find k^* is clear: First, we use either Seq-Sampler or Mul-Sampler to obtain estimates K_a and K_r (of k_a^* and k_r^*, respectively). Then computing $\min(K_a, K_r)$ gives us the approximation of k^* that we need.

5.2   Seq-Sampler

Seq-Sampler aims to compute k_a^* and k_r^* by following a 'purely sequential' approach, described in Sect. 2.4. We compute k_a^* by the approximation [11,31]:

$$K_a^{Seq} = \text{smallest integer } k \geq k_0 \text{ s.t. } S_k^2 z_{\alpha/2}^2 / \delta^2. \tag{11}$$

The approximation of k_r^* is similar:

$$K_r^{Seq} = \text{smallest integer } k \geq k_0 \text{ s.t. } S_k^2 z_{\alpha/2}^2 / \left(\delta \left|\bar{Z}_k\right|\right)^2. \tag{12}$$

As discussed, $K^{Seq} = \min(K_a^{Seq}, K_r^{Seq})$ is the resulting approximation of k^*. Alternatively, we can mix the conditions in (11) and (12) to write: K^{Seq} is the smallest integer $k \geq k_0$ such that

$$S_k^2 \left(z_{\alpha/2}/\delta\right)^2 \times \min\left(1, \left(1/\bar{Z}_k\right)^2\right). \tag{13}$$

To find K_{Seq}, Seq-Sampler approximates σ^2 and μ successively in an *incremental* manner, while continually checking if the condition (13) is satisfied.

Seq-Sampler is presented in Algorithm 2. First, we sample k_0 blocks $(B_1, \ldots, B_{k_0})$, and use them to obtain the estimates $(Z_1, \ldots, Z_{k_0})$. We then check the condition (13) with $S_{k_0}^2$ and $\bar{Z}_{k_0}$. If it is satisfied, we stop the algorithm with $K^{Seq} = k_0$.

Next, we keep sampling an additional block (while computing the corresponding estimate) *one at a time*. We approximate σ^2 and μ by the sample variance and sample mean, respectively, on the estimates we've obtained. For example, suppose we have obtained the list $(Z_1, \ldots, Z_k)$. The next step is to sample B_{k+1} and obtain Z_{k+1}, then we use S_{k+1}^2 for σ^2 and $\bar{Z}_{k+1}$ for μ. This iterative process allows us to evaluate the expression (13) at every step. The process halts when we have obtained *enough* blocks, i.e., when the number of blocks exceeds the estimate of (13).

Algorithm 2. The Seq-Sampler Algorithm

Input: Dataset X, Statistics T, Accuracy parameter δ, Confidence parameter α
Output: K^{Seq} and $Z_1, \ldots, Z_{K^{Seq}}$

1 Set $k_0 = 10$.
2 Set $\eta = 1/10$ and block size $b = (1/\delta)^{1+\eta}$.
3 If $b \times k_0 > \text{size}(X)$, output -1 and $[]$.
4 Set list of estimates $Z = []$.
5 **for** $i = 1$ *to* k_0 **do**
6 $\quad$ Sample block B_i (of size b).
7 $\quad$ Add estimate $Z_i = T(B_i)$ to Z.
8 **end**
9 Set $k = k_0$.
10 **while** $k < (S_k^2 z_{\alpha/2}^2 / \delta^2) \times \min\left(1, \left(1/\bar{Z}_k\right)^2\right)$ **do**
11 $\quad$ Sample block B_{k+1} (of size b).
12 $\quad$ If not enough data, output -1 and $[]$.
13 $\quad$ Otherwise, add estimate $Z_{k+1} = T(B_{k+1})$ to Z.
14 $\quad$ Increment k.
15 **end**
16 Output $K^{Seq} = k$ and $Z_1, \ldots, Z_{K^{Seq}}$.

Like [11], we need $k \geq k_0$, where k_0 is the *mininum* number of blocks. As discussed in [35], k_0 should not be too small. If, for instance, $k = 2$, the two estimates Z_1 and Z_2 may be similar such that $S_{k_0}^2$ is close to 0, causing early stopping. In Algorithm 2, we choose $k_0 = 10$ like [35], since it is unlikely that we accidentally encounter 10 similar estimates.

The algorithm might operate incorrectly if there were not enough data to work on. This situation may happen during sampling of consecutive blocks (Line 12), or even with the initial k_0 blocks (Line 3). In such cases, we output $K^{Seq} = -1$ and $Z = []$.

5.3 Mul-Sampler

Simpler than Mul-Sampler, Mul-Sampler computes k_a^* and k_r^* by following a 'multi-stage' approach, also described in Sect. 2.4. The idea of multi-stage methods is to sample some *pilot* data, and use it to approximate unknown parameters like [31].

Details of Mul-Sampler are described as follows. First, we compute k_0 estimates $(Z_1, \ldots, Z_{k_0})$, based on which we use $S_{k_0}^2$ and $\bar{Z}_{k_0}$ in place of σ^2 and μ, respectively. k_a^* is approximated by

$$K_a^{Mul} = \text{smallest integer } k \geq \max\left\{k_0, S_{k_0}^2 z_{\alpha/2}^2 / \delta^2\right\}. \tag{14}$$

The strategy is straightforward: If the rightmost term in (14) is less than k_0, we set $K_a^{Mul} = k_0$. Otherwise, $K_a^{Mul} = S_{k_0}^2 z_{\alpha/2}^2 / \delta^2$, and we sample the

---

Algorithm 3. The Mul-Sampler Algorithm

Input: Dataset X, Statistics T, Accuracy parameter δ, Confidence parameter α
Output: K^{Mul} and $Z_1, \ldots, Z_{K^{Mul}}$

1 Set $\gamma = 0.5$ and $k_0 = \max \left\{ 2, \left\lceil (z_{\alpha/2}/\delta)^{2/(1+\gamma)} \right\rceil \right\}$.
2 Set $\eta = 1/10$ and block size $b = (1/\delta)^{1+\eta}$.
3 If $b \times k_0 > \mathsf{size}(X)$, output -1 and $[]$.
4 Set list of estimates $Z = []$.
5 **for** $i = 1$ to k_0 **do**
6 $\quad$ Sample block B_i (of size b).
7 $\quad$ Add estimate $Z_i = T(B_i)$ to Z.
8 **end**
9 Set $K^{Mul} = $ smallest integer $k \geq \max \left\{ k_0, S_{k_0}^2 z_{\alpha/2}^2 / \delta^2 \times \min \left(1, (1/\bar{Z}_{k_0})^2 \right) \right\}$.
10 If $b \times K^{Mul} > \mathsf{size}(X)$, output -1 and $[]$.
11 **if** $K_{mul} > k_0$ **then**
12 $\quad$ **for** $i = k_0 + 1$ to K^{Mul} **do**
13 $\quad\quad$ Sample block B_i (of size b).
14 $\quad\quad$ Add estimate $Z_i = T(B_i)$ to Z.
15 $\quad$ **end**
16 **end**
17 Output K^{Mul} and $Z_1, \ldots, Z_{K^{Mul}}$.

---

difference (i.e., the blocks $B_{k_0+1}, \ldots, B_{K_{Mul}}$). We have a similar formula for the approximation of k_r^*:

$$K_r^{Mul} = \text{smallest integer } k \geq \max \left\{ k_0, S_{k_0}^2 z_{\alpha/2}^2 / \left(\delta | \bar{Z}_{k_0} | \right)^2 \right\}, \qquad (15)$$

and the strategy is exactly the same. Again, (14) and (15) can be combined together to form the following approximation: $K^{Mul} = $ smallest integer

$$k \geq \max \left\{ k_0, S_{k_0}^2 \left(z_{\alpha/2}/\delta \right)^2 \times \min \left(1, \left(1/\bar{Z}_{k_0} \right)^2 \right) \right\}, \qquad (16)$$

The whole procedure is given in Algorithm 3. Note that k_0 is no longer the minimum number of blocks as in Seq-BPMiner. Instead, we consider k_0 the *pilot size*, i.e., size of the pilot data that we use to approximate σ^2 and μ. For this reason, the choice of k_0 decides the quality of our approximations, and therefore requires some forethought. In this paper, we use the pilot size

$$k_0 = \max \left\{ 2, \left\lceil (z_{\alpha/2}/\delta)^{2/(1+\gamma)} \right\rceil \right\} \qquad (17)$$

that is proved in [30,31] to ensure good asymptotic results. Here $\gamma > 0$ is a fixed number. In this paper, we let $\gamma = 0.5$.

Similar to Seq-Sampler, there exist cases in which the data X is not large enough. For example, the pilot size k_0 (Line 3) or the final number of blocks K^{Mul} (Line 10) may require more data than what we have in hand. Again, we handle such cases by outputting $K^{Mul} = -1$ and $Z = []$.

Algorithm 4. The ManageBudget Procedure

Input: Dataset X, Effective estimates $(Z_1, \dots, Z_K)$
Output: Updated X (with unexhausted budgets)

1 **foreach** *Data point X_j in X* **do**
2 | **if** *X_j belongs to any block B_i whose resulting estimate Z_i is in $(Z_1, \dots, Z_K)$* **then**
3 | | PrivacyBudget(X_j) = PrivacyBudget$(X_j) - \epsilon$.
4 | **end**
5 | Remove X_j from X if PrivacyBudget$(X_j) < \epsilon$.
6 **end**

5.4 ManageBudget

Before embarking on the main algorithm, we need the final missing piece: a procedure to handle all personal privacy budgets (in the spirit described in Sect. 4.4). Algorithm 4 sets out to accomplish this. The algorithm takes as input two arguments: (1) the dataset X and (2) the estimates $(Z_1, \dots, Z_K)$ that we've obtained (from either Seq-Sampler or Mul-Sampler). If a data point is involved in the computation, we subtract the privacy guarantee ϵ from its budget. Subsequently, we get rid of the data points whose budgets have run out.

5.5 Plugging All Pieces Together

We are now in the position to blend together all elements: sequential estimation (Seq-Sampler/Mul-Sampler), SaG, and personal privacy budget (ManageBudget). The main solution, BPMiner, is presented in Algorithm 5.

Line 3 of the algorithm applies either Seq-Sampler (Algorithm 2) or Mul-Sampler (Algorithm 3) to extract the list of estimates $Z_1, \dots, Z_K$. As mentioned, K is a general notation for both K^{Seq} and K^{Mul}. Meanwhile, Line 7 adds differential privacy by applying the 'aggregate' part of the SaG framework on $Z_1, \dots, Z_K$. In this paper, we use the method of *Widened Winsorized Mean*, which is based on differentially private quantiles (see both in [38]). This SaG method has been proved to converge to the same distribution as the non-private estimator [38]. Finally, Line 8 of Algorithm 5 is responsible for handling personal privacy budgets. Specifically, it applies the ManageBudget procedure (Algorithm 4) on the list of estimates $(Z_1, \dots, Z_K)$ we've obtained from Line 3.

Line 4 checks whether execution of Seq-Sampler/Mul-Sampler results in $K = -1$, signalling insufficient data for this particular computation. If this is true, we do not terminate the loop, since the data in hand may still be useful for another computation. Therefore, we output Null and skip to the next iteration, with the privacy budgets untouched. The loop only terminates when we have executed all required computations.

Algorithm 5. The BPMiner Algorithm

Input: Dataset $X = (X_1, \ldots, X_n)$, Privacy parameter ϵ, Accuracy parameter δ,
Confidence parameter α, List of computations $T_1, \ldots, T_M$
Output: List of estimates $\hat{\theta}_T^{(1)}, \ldots, \hat{\theta}_T^{(M)}$

1   Set $i = 1$.
2   **while** $i \leq M$ **do**
3   Execute Seq-Sampler/Mul-Sampler on X, T_i, δ and α to obtain K and
     $(Z_1, \ldots, Z_K)$.
4   **if** $K \neq -1$ **then**
5   Compute $\hat{\theta}_T^{(i)}$ by 'aggregating' on $(Z_1, \ldots, Z_K)$.
6   Execute ManageBudget on X and $(Z_1, \ldots, Z_K)$.
7   **else**
8   Set $\hat{\theta}_T^{(i)} = $ Null.
9   **end**
10   Increment i.
11 **end**
12 Output $\hat{\theta}_T^{(1)}, \ldots, \hat{\theta}_T^{(M)}$.

5.6 Multivariate Setting

The discussion so far works well for *univariate* (one-dimensional) statistics, e.g., the mean value of some attribute. Extension of BPMiner to multivariate estimators, however, requires some thought.

The problem lies in σ^2, which is replaced by the *covariance matrice* in the multivariate setting. There are certain difficulties in working with covariance matrices. First, estimation of covariance matrices is not as straightforward as computing σ^2 [7]. Second, evaluating high-dimensional covariance matrices (especially at multiple steps as in Seq-Sampler) introduces high computational burden, disturbing the computational efficiency we aim to gain. Finally, sequential estimation in multivariate statistics is much more sophisticated than the univariate counterpart [4,35], making it harder to be applied in our case.

To sidestep these issues, we require (δ, α)-accuracy simultaneously for all dimensions, that is,

$$\Pr\left(\cap_{i=1}^d |\hat{\theta}_{T|i} - \theta_i| \leq \max(\delta, \delta|\theta_i|) \right) \geq 1 - \alpha. \tag{18}$$

Such requirement allows us to work independently for each dimension. Note that the expression in parentheses equals $\cap_{i=1}^d \hat{\theta}_{T|i} \in [\theta_i \pm \max(\delta, \delta|\theta_i|)]$. In other words, we relax the general (spherical/ellipsoidal) intervals and work with axis-parallel rectangles instead. Nevertheless, experimental results will show later that this assumption works reasonably well in multivariate cases.

Suppose we have ensured (δ, α)-accuracy of $\hat{\theta}_T$ for every dimension, that is, for $i = 1, \ldots, d$, $\Pr(|\hat{\theta}_{T|i} - \theta_i| \leq \max(\delta, \delta|\theta_i|)) \geq 1 - \alpha$. Recall Bonferroni inequalities: for events $A_1, \ldots, A_p$, we have $\Pr(\cap_{i=1}^p A_i) \geq 1 - \sum_{i=1}^p \Pr(\bar{A}_i)$. Therefore, the left-hand side of (18) is greater than or equal to

$$1 - \sum_{i=1}^{d} \Pr\left(|\hat{\theta}_{T|i} - \theta_i| > \max(\delta, \delta|\theta_i|)\right) \geq 1 - \sum_{i=1}^{d} \alpha = 1 - d\alpha.$$

Therefore, it turns out that we can ensure (δ, α)-accuracy for $\hat{\theta}_T$ by making sure each marginal value $\hat{\theta}_{T|i}$, $i = 1, \ldots, d$, is $(\delta, \alpha/d)$-accurate. The whole argument can also be seen in [35].

To facilitate this idea, only minor changes to the previous algorithms are required. The BPMiner algorithm is unchanged. In the Seq-Sampler algorithm, we change α to α/d, and replace the condition in Line 10 with

$$\text{for all } i = 1, \ldots, d: \quad k < (S_{k|i}^2 z_{\alpha/2}^2/\delta^2) \times \min\left(1, (1/\bar{Z}_{k|i})^2\right),$$

where $S_{k|i}$ and $\bar{Z}_{k|i}$ are respectively the sample variance and sample mean of $(Z_1, \ldots, Z_k)$ along the i-th dimension. Similarly, in the Mul-Sampler algorithm, we replace Line 9 with

$$K^{Mul} = \max_{i}\{k_i\} \text{ where } k_i = \text{smallest integer}$$
$$\geq \max\left\{k_0, S_{k_0|i}^2 z_{\alpha/2}^2/\delta^2 \times \min\left(1, (1/\bar{Z}_{k_0|i})^2\right)\right\}.$$

6   Analysis and Evaluation

6.1   Analysis

We provide rigorous proof of why BPMiner satisfies our requirements of privacy and accuracy (as specified in Sect. 3).

Privacy For a Single Computation. Using sequential estimation, our approach aims to make efficient use of data. Once the optimal size K is obtained, we apply SaG on $Z_1, \ldots, Z_K$. As a consequence, whether our algorithm preserves privacy depends completely on SaG methods. Since such techniques are proposed as differentially private mechanisms [18,29,37,38], it easily follows that the proposed algorithm also preserves ϵ-differential privacy.

For Multiple Computations. Suppose we need to execute M computations, and we want to ensure ϵ-differential privacy under k-fold adaptive composition (as motivated in Sect. 3). We need a preliminary result from [17].

Lemma 3. *For every $\epsilon > 0$, the family of ϵ-differentially private algorithms satisfies $k\epsilon$-differential privacy under k-fold adaptive composition.*

Suppose an individual's data has been used in M different samples in correspondence to the M computations. We previously claimed that our algorithm is ϵ-differential private on handling each computation. By Lemma 3, our approach also provides $M\epsilon$-differential privacy under M-fold adaptive composition. This essentially makes sure that an adversary cannot tell whether an individual has participated in any of the M computations.

Accuracy. Here we present a result about the asymptotic accuracy of our solution. First, let us recall some notation from Sect. 5: K_a denotes either K_a^{Seq} or K_a^{Mul}, and K_r represents either K_r^{Seq} or K_r^{Mul}. Some preliminary results of K_a and K_r are given as follows.

Lemma 4. $K_a = K_a(\delta)$ and $K_r = K_r(\delta)$ are both functions of δ, and

1. When $\delta \to 0$, $K_a(\delta) \to \infty$ with probability 1 and also $K_r(\delta) \to \infty$ with probability 1,
2. $(K_a\delta^2)/(\sigma^2 z_{\alpha/2}^2) \to 1$ almost surely,
3. $(K_r\mu^2\delta^2)/(\sigma^2 z_{\alpha/2}^2) \to 1$ almost surely.

Proof of the results for K_a is standard in sequential estimation (see [11] for K_a^{Seq} and [31] for K_a^{Mul}). Properties of K_r are simply trivial modifications from those results (see [32]). We omit these proof due to the page constraint. We are now ready to introduce our main results in Theorem 3. The last result of this theorem firmly states that the output of BPMiner is (δ, α)-accurate.

Theorem 3. Suppose $\hat{\theta}_{T|a}(X)$ and $\hat{\theta}_{T|r}(X)$ are the estimators constructed according to Algorithm 5 for absolute (δ, α) accuracy and relative (δ, α) accuracy, respectively. Also, recall that $\hat{\theta}_T(X)$ is the output of Algorithm 5 for (δ, α) accuracy. Then as $\delta \to 0$, we have the following results:

1. $\Pr(|\hat{\theta}_{T|a}(X) - \theta| \le \delta) \to 1 - \alpha$,
2. $\Pr(|\hat{\theta}_{T|r}(X) - \theta| \le \delta|\theta|) \to 1 - \alpha$,
3. $\Pr(|\hat{\theta}_T(X) - \theta| \le \max(\delta, \delta|\theta|)) \to 1 - \alpha$.

Proof. We consider the first part regarding absolute (δ, α)-accuracy. When $\delta \to 0$, $K_a = K_a(\delta) \to \infty$ with probability 1 by the first result in Lemma 4. Then by Theorem 1, when $\delta \to 0$, we can consider the prescribed level of accuracy on $\bar{Z}$ (rather than $\hat{\theta}_{T|a}(X)$), i.e., $\Pr(|\hat{\theta}_{T|a}(X) - \theta| \le \delta) \to \Pr(|\bar{Z} - \theta| \le \delta) = \Pr(\sqrt{K_a}|\bar{Z} - \theta|/\sigma \le \delta\sqrt{K_a}/\sigma)$. As previously derived in Sect. 5, this probability equals

$$\Pr\left(\left|\sqrt{K_a}\frac{|\bar{Z} - \mu|}{\sigma} \pm \sqrt{K_a}\frac{|\theta - \mu|}{\sigma}\right| \le \frac{\delta\sqrt{K_a}}{\sigma}\right). \tag{19}$$

We analyze the term $\sqrt{K_a}|\theta - \mu|/\sigma$. As $Z_1, \ldots, Z_{K_a}$ are i.i.d., $\mu = \mathbb{E}Z_1 = \theta + O(1/b)$ by the linear-bias assumption. Therefore,

$$\sqrt{K_a}\frac{|\theta - \mu|}{\sigma} = \frac{\delta\sqrt{K_a}}{z_{\alpha/2}\sigma} \times \frac{z_{\alpha/2}}{\delta} \times |\theta - \mu|.$$

In the right-hand side, the first term tends to 1 as $\delta \to 0$ by the second result in Lemma 4. The third term is $O(1/b)$, while the second term is $o(b)$ since we have chosen b such that $1/\delta = o(b)$. Putting them together gives us $\sqrt{K_a}|\theta - \mu|/\sigma = o(b) \times O(1/b) = o(1)$, so that it tends to 0 as $\delta \to 0$. Therefore, with this choice of b, the second summand in (19) vanishes asymptotically as $d \to 0$.

Meanwhile, it has been known from [3] that the CLT applies to a random number of summands, that is,

$$\sqrt{K_a}(\bar{Z} - \mu)/\sigma \xrightarrow{D} N(0,1). \tag{20}$$

In addition, by the second result of Lemma 4, the right-hand side in (19), $\delta\sqrt{K_a}/\sigma$, tends to $z_{\alpha/2}$. Putting this together with (20), we come up with $\lim_{\delta \to 0} \Pr(\sqrt{K_a}|\bar{Z} - \mu|/\sigma \le \delta\sqrt{K_a}/\sigma) = 1 - \alpha$. Combining this with the fact that the second summand in (19) vanishes completes the proof for the first result of Theorem 3.

The second result can be proved similarly, with K_r in place of K_a. A notable difference is when (19) is replaced by

$$\Pr\left(\left|\sqrt{K_r}\frac{|\bar{Z} - \mu|}{\sigma} \pm \sqrt{K_r}\frac{|\theta - \mu|}{\sigma}\right| \le \frac{|\theta|\delta\sqrt{K_r}}{\sigma}\right). \tag{21}$$

The left-hand side of (21) can be processed exactly like before. The right-hand side can be rewritten as $|\theta\frac{\delta\sqrt{K_r}}{\sigma}| = |\mu\frac{\delta\sqrt{K_r}}{\sigma} - \frac{\delta\sqrt{K_r}}{\sigma}O(1/b)|$. We write the second summand as

$$\frac{|\mu|\delta\sqrt{K_r}}{z_{\alpha/2}\sigma} \times \frac{z_{\alpha/2}}{|\mu|} \times \frac{1}{\delta} \times O(1/b).$$

The first term tends to 1 as $\delta \to 0$ by the third result in Lemma 4. With the same choice of b as before, the third term is $o(b)$. Overall, the whole summand tends to 0 as $\delta \to 0$, and the right-hand side of (21) is only left with $|\mu|\delta\sqrt{K_r}/\sigma$. Now we can proceed the proof exactly like before (with help of the third result in Lemma 4) to obtain the desired result.

The third result in Theorem 3 involves mixed (δ, α) accuracy and $K = \min(K_a, K_r)$. It trivially follows from the first two results (see [32]).

6.2   Evaluation

We now give some empirical evaluation of our algorithms. In the upcoming experiments, algorithm implementations are written in Matlab, and all computations are performed on a 6-core Intel Xeon E5–1650 running at 3.20GHz with 16GB of RAM.

Experimental Setup Estimators / Statistics. We evaluate the performance of our algorithms using three different estimators: (1) *mean* (Mean), (2) *linear regression* (LinReg), and (3) *logistic regression* (LogReg). These statistics are generically asymptotically normal [38], and commonly seen in the data mining and statistics community.

Applying SaG algorithms (Algorithm 1) requires specifying the output range. Since this range affects the additive noise for differential privacy, a poor choice may lead to undesirable accuracy. It is therefore important to use the same output range during comparison of algorithms, and we propose the output range

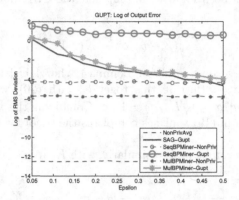

(a) Accuracy of Mean estimator (GUPT-perturbed) on dataset XMul-Gauss

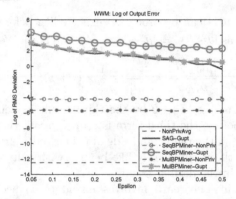

(b) Accuracy of Mean estimator (WWM-perturbed) on dataset XMulGauss

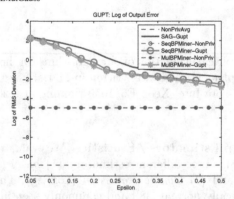

(c) Accuracy of Mean estimator (GUPT-perturbed) on dataset YearPredictionMSD

Fig. 1. Selected results for output accuracy

of each estimator as follows. For the mean estimator, the output range is taken as the min-max range of the data. For linear regression and logistic regression, the output range is set to be $[-10, 10]$. Rather than applying black-box implementations, here we use constrained optimization with coefficients bounded in $[-10, 10]$.

Finally, these statistics are different in that they take different amounts of time to compute. Approximating the mean requires only one scan of data, and is therefore pretty fast. Meanwhile, computing the linear and logistic regression weights (especially with constrained optimization) may take some time for massive datasets.

Datasets. We experiment with 4 datasets. The first three, so-called XMul-Gauss, XMulGaussYLin and XMulGaussYBernLin, are synthetic datasets with $n = 1,000,000$ rows and $d = 10$ columns in size. With XMulGauss, each row is generated as $X \sim \text{Normal}(0, I_d)$. With XMulGaussYLin, the first $d-1$ columns are generated as $X \sim \text{Normal}(0, I_{d-1})$, and the last column is generated according to the linear model $Y = X_i^T \mathbf{1}_{d-1} + \epsilon$ with $\epsilon \sim \text{Normal}(0, 1)$. The dataset XMulGaussYBernLin is essentially the same as XMulGaussYLin, except that the last column is $Y \sim \text{Bernoulli}((1 + \exp(X_i^T \mathbf{1}_{d-1}))^{-1})$. Here I_d is the $d \times d$ identity matrix, and $\mathbf{1}_d$ the d-dimensional vector of all ones. We note that such data generation is almost identical to [22].

The last one is a real-world dataset named YearPredictionMSD. According to [20], this is the largest dataset for regression in the UCI Machine Learning Repository [5]. It is a subset of the *Million Song* dataset [6]. The dataset consists of $n = 515345$ rows and $d = 91$ columns. There are 90 features measuring the timbre average and covariance values. The target column to be predicted is the year, taking values from 1922 to 2011. Like [20], we linearly scale the year to $[0, 1]$. In addition, each feature column is normalized by its mean and standard deviation.

The mean estimator is experimented over the datasets XMulGauss and YearPredictionMSD. Linear regression is applied on XMulGaussYLin and YearPredictionMSD, while logistic regression is tested over XMulGaussYBernLin.

Algorithms. We compare between 3 algorithms: SaG, Seq-BPMiner and Mul-BPMiner. The SaG algorithm is considered the baseline. Seq-BPMiner and Mul-BPMiner are variations of BPMiner that use Seq-Sampler and Mul-Sampler, respectively. All of the 3 algorithms have been proved to be differentially private. In addition, note that there have been some suggestions regarding the noise added for differential privacy. In this paper, we use two methods: GUPT (the GUPT-Tight method in [29]) and WWM (proposed in [38]). For each test case (e.g., the dataset XMulGauss and the Mean computation), we execute both GUPT and WWM versions.

Results Output Accuracy. The mentioned algorithms preserve differential privacy by adding a sufficient amount of noise computed based on ϵ. With small ϵ, more privacy is required, so more noise is added and the output is less accurate. Meanwhile, larger ϵ means lighter privacy guarantee, allowing the output to be more accurate.

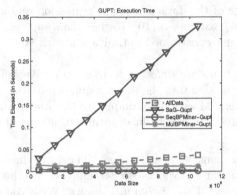

(a) Execution time of Mean estimator (GUPT-perturbed) on dataset XMulGauss

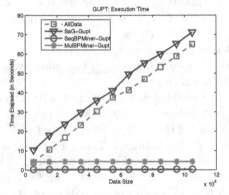

(b) Execution time of Linear Regression estimator (GUPT-perturbed) on dataset XMulGaussYLin

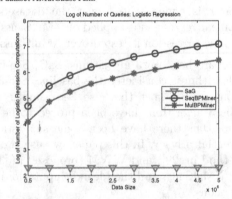

(c) Number of Logistic Regression computations on dataset XMulGaussYBernLin

Fig. 2. Selected results for computational and budget efficiency

We measure the accuracy of outputs for each algorithm over a range of ϵ from $\epsilon = 0.05$ to $\epsilon = 0.5$. The output accuracy of an algorithm is gauged by the RMS deviation between that output and the *'true'* output, i.e., one that is obtained when the estimator is used over the complete dataset. We only show selected results in Fig. 1, though other combinations yield qualitatively analogous outcome. Figures 1(a) and (b) show the GUPT and WWM output accuracy for the dataset XMulGauss and the Mean computation, respectively. Figure 1(c) illustrates the GUPT accuracy for the real dataset YearPredictionMSD, also with the Mean computation.

Besides the fact that the relationship between ϵ and output accuracy is well reflected in these figures, there are two interesting observations we can make from the plots. First, note that the performance of Seq-BPMiner and Mul-BPMiner cannot be better than SaG since they use less data for computation. This is illustrated in all three plots where the non-private outputs of Seq-BPMiner and Mul-BPMiner has higher error than SaG. The story changes, however, when noise is added to the output. As we can see from the results, Mul-BPMiner have (almost) comparable performance to the baseline SaG algorithm.

Second, as mentioned in [31], Mul-Sampler tends to *over-sample* (i.e., giving approximations larger than the real k^*) while Seq-Sampler has second-order asymptotic properties and provides more accurate approximations. Therefore, the number of blocks produced by Seq-Sampler is less than that of Mul-Sampler. For both GUPT and WWM noise-adding methods, the variance of the noise is inversely proportional to the number of blocks. This explains the worse performance of Seq-BPMiner in comparison to Mul-BPMiner.

Computational Efficiency. Figures 2(a) and (b) show the computational efficiency of four algorithms (including the AllData algorithm where the computation is applied over the complete dataset). In these experiments, we vary the size of dataset from $n = 550,000$ to $n = 10,500,000$ (the number of columns is still fixed as $d = 10$), and measure the computational efficiency of each algorithm by its execution time. Therefore, these sets of experiments are not applicable to the fixed-sized real-world dataset YearPredictionMSD. Results on logistic regression are similar to linear regression, and therefore omitted due to the space constraint. A few observations can be made here. First, a common observation between the two plots is that both Seq-BPMiner and Mul-BPMiner have almost constant time *regardless of* the increase in the data size. This is understandable since in our derivation, the (approximated) optimal number of blocks (produced by either Seq-Sampler or Mul-Sampler) does not depend on the size of the original dataset. Meanwhile, AllData and SaG performs computationally worse as expected for increasing data size.

Second, regarding the mean computation, in small data regimes (say, less than 2,000,000 rows), both Seq-BPMiner and Mul-BPMiner are worse than the AllData algorithm. The reason is that the mean is relatively fast to compute (especially in Matlab), while Seq-BPMiner and Mul-BPMiner have overhead time of moving between data blocks. For large data regimes, the time to compute the mean becomes more significant and soon dominates the overhead time, resulting in Seq-BPMiner and Mul-BPMiner being more efficient than SaG. On the other hand, regarding linear/logistic regression, Seq-BPMiner and Mul-BPMiner are always better than SaG due to the long execution time of linear/logistic regression computations on large datasets. The overhead time is now ignorable compared to the actual execution time on data blocks.

The final observation is that SaG always performs the worst. This can be explained by the moderate size of each data block (compared to Seq-BPMiner and Mul-BPMiner) and the overhead of moving between the blocks.

Budget Efficiency. We evaluate the budget efficiency of the algorithms using the number of computations they allow before budget exhaustion. Figure 2(c) shows the number of logistic regression computations when we increase the size of the dataset XMulGaussYBernLin from $n = 550,000$ to $n = 5,500,000$. The privacy budget for the SaG algorithm is set to 1. For a fair comparison, the personal privacy budget for each data instance is also 1. The privacy guarantee for each logistic regression query is $\epsilon = 0.1$. Again, results for the mean and linear regression computations (omitted here) are qualitatively analogous.

Two observations can be drawn. First, the number of computations by the SaG algorithm is always constant (10 queries in our setting) regardless of the increase in data size. This is because the SaG uses up data per computation for all sizes of the dataset. Meanwhile, Seq-BPMiner and Mul-BPMiner operate on samples of data only, so there exists an increase in the number of computations when more data is available. Second, as previously mentioned, Seq-BPMiner uses a smaller number of blocks than Mul-BPMiner, which means less data is consumed by Seq-BPMiner per computation. With multiple computations, it is understandable that Seq-BPMiner allows more queries to be executed before the privacy budgets are exhausted.

7   Conclusion

BPMiner offers a theoretically strong solution for the problem of private analysis on large-scale datasets. It is differentially private per computation, while for M computations, it provides $M\epsilon$-differential privacy under M-fold adaptive composition. Rigorous asymptotic analysis demonstrates that though only a (much) smaller sample is used, for large data regimes (more precisely, large number of blocks), the accuracy of BPMiner is well-controlled with prescribed accuracy. Finally, the challenges of computational and budget efficiency are simultaneously addressed by the use of sequential estimation. Experimental results have shown that BPMiner significantly outperforms the SaG algorithms in term of the execution time and number of queries given a budget constraint, while having comparable accuracy on many datasets. With these characteristics, we believe BPMiner makes an ideal solution for private analysis on humongous data volumes.

Regarding possible future work, one might want to consider a hybrid approach that adaptively combines SaG (addressing small data regimes) and

BPMiner (addressing large data regimes). Alternatively, non-asymptotic versions of sequential estimation [10] can be employed, since non-asymptotic analysis provides guarantees for any (even small) sample size. However, this is highly non-trivial as accordingly, one would also need finite-sample analysis of SaG algorithms.

Acknowledgement. This work was funded by A*Star Science and Engineering Research Council (SERC)'s Thematic Strategic Research Programme (TSRP) grant number 102 158 0038.

A   Proof of Theorem 1

For estimators from [18,29,37], the Laplace noise Y is added directly to $\bar{Z}$ as in Algorithm 1. We have the following useful lemma.

Lemma 5. *Suppose the Laplace noise is given by* $Y = Y_k = \text{Laplace}(\text{Range}/k\epsilon)$. *Then* $Y_k \xrightarrow{P} 0$ *as* $k \to \infty$.

Proof (of Lemma 5). Given $\epsilon > 0$, consider the probability $\Pr(|Y_k - 0| < \epsilon)$. This probability equals

$$\Pr(|Y_k| < \epsilon) = \Pr(-\epsilon < Y_k < \epsilon) = F_{Y_k}(\epsilon) - F_{Y_k}(-\epsilon).$$

The CDF of a Laplace random variable X is given by $F_X(x) = \frac{1}{2}\exp\left(\frac{x-\mu}{\sigma}\right)$ if $x < \mu$, and $F_X(x) = 1 - \frac{1}{2}\exp\left(-\frac{x-\mu}{\sigma}\right)$ otherwise. Since Y_k has a Laplace distribution with parameters $\mu = 0$ and $\sigma = \text{Range}/(k\epsilon)$, computing $F_{Y_k}(\epsilon)$ and $F_{Y_k}(-\epsilon)$ (with $\epsilon > 0 = \mu$) gives us

$$\Pr(|Y_k - \mu| < \epsilon) = F_{Y_k}(\epsilon) - F_{Y_k}(-\epsilon) = 1 - \exp\left(-\frac{k\epsilon^2}{\text{Range}}\right) \to 1$$

as $k \to \infty$. Therefore, it follows from the definition of convergence in probability that $Y_k \xrightarrow{P} 0$ as $k \to \infty$.

Using Lemma 5, the proof is straightforward for the statistics proposed in [18,29,37]: $\Pr(|\hat{\theta}_T(X) - \theta| \leq \delta) = \Pr(|\bar{Z} + Y_k - \theta| \leq \delta) \to \Pr(|\bar{Z} - \theta| \leq \delta)$.

For the estimator from [38], the noise Y is added to the winsorized mean rather than to $\bar{Z}$, so the above argument does not apply. Nevertheless, the results extracted from [38] are particularly useful in deriving the proof. First, we observe that

$$\Pr(|\hat{\theta}_T(X) - \theta| \leq \delta) = \Pr(\theta - \delta \leq \hat{\theta}_T(X) \leq \theta + \delta)$$
$$= F_{\hat{\theta}_T}(\theta + \delta) - F_{\hat{\theta}_T}(\theta - \delta), \tag{22}$$

where F_X denotes the CDF of X. By Corollary 10 in [38], the KS distance between $\hat{\theta}_T$ and $\bar{Z}$ goes to 0 as k goes to infinity. This leads to the fact that $\hat{\theta}_T$

converges in distribution to $\bar{Z}$. In other words, when $k \to \infty$, $F_{\hat{\theta}_T}(t) \to F_{\bar{Z}}(t)$, and the expression (22) converges to

$$F_{\bar{Z}}(\theta + \delta) - F_{\bar{Z}}(\theta - \delta) = \Pr(\theta - \delta \leq \bar{Z} \leq \theta + \delta)$$
$$= \Pr(|\bar{Z} - \theta| \leq \delta),$$

which completes the proof.

References

1. Agarwal, A., Chapelle, O., Dudík, M., Langford, J.: A reliable effective terascale linear learning system. J. Mach. Learn. Res. **15**(1), 1111–1133 (2014)
2. Agarwal, S., Milner, H., Kleiner, A., Talwalkar, A., Jordan, M.I., Madden, S., Mozafari, B., Stoica, I.: Knowing when you're wrong: building fast and reliable approximate query processing systems. In: International Conference on Management of Data, SIGMOD 2014, Snowbird, UT, USA, 22–27 June 2014, pp. 481–492 (2014)
3. Anscombe, F.J.: Large-sample theory of sequential estimation. Math. Proc. Cambridge Philos. Soc. **48**(4), 600–607 (1952)
4. Aoshima, M., Yata, K.: Two-stage procedures for high-dimensional data. Sequential Analysis **30**(4), 356–399 (2011)
5. Bache, K., Lichman, M.: UCI Machine Learning Repository. IOS Press, Amsterdam (2013)
6. Bertin-Mahieux, T., Ellis, D.P.W., Whitman, B., Lamere, P.: The million song dataset. In: Proceedings of the 12th International Society for Music Information Retrieval Conference ISMIR 2011, Miami, Florida, USA, 24–28 October 2011, pp. 591–596 (2011)
7. Bickel, P.J., Levina, E.: Regularized estimation of large covariance matrices. Ann. Stat. **36**, 199–227 (2008)
8. Blum, A., Dwork, C., McSherry, F., Nissim, F.: Practical privacy: the sulq framework. In: Proceedings of the Twenty-fourth ACM SIGACT-SIGMOD-SIGART Symposium on Principles of Database Systems, Baltimore, Maryland, USA, 13–15 June 2005, pp. 128–138 (2005)
9. Cai, Z., Gao, Z.J., Luo, S., Perez, L.L., Vagena, Z., Jermaine, C.M.: A comparison of platforms for implementing and running very large scale machine learning algorithms. In: International Conference on Management of Data, SIGMOD 2014, Snowbird, UT, USA, 22–27 June 2014, pp. 1371–1382 (2014)
10. Chen, J., Chen, X.: A new method for adaptive sequential sampling for learning and parameter estimation. In: Kryszkiewicz, M., Rybinski, H., Skowron, A., Raś, Z.W. (eds.) ISMIS 2011. LNCS, vol. 6804, pp. 220–229. Springer, Heidelberg (2011)
11. Chow, Y.S., Robbins, H.: On the asymptotic theory of fixed-width sequential confidence intervals for the mean. Ann. Math. Stat. **36**(2), 457–462 (1965)
12. Condie, T., Mineiro, P., Polyzotis, N., Weimer, M.: Machine learning for big data. In: Proceedings of the ACM SIGMOD International Conference on Management of Data, SIGMOD 2013, New York, NY, USA, 22–27 June 2013, pp. 939–942 (2013)
13. Condie, T., Mineiro, P., Polyzotis, N., Weimer, M.: Machine learning on big data. In: 29th IEEE International Conference on Data Engineering, ICDE 2013, Brisbane, Australia 8–12 April 2013, pp. 1242–1244 (2013)

14. Dwork, C.: Differential privacy. In: Bugliesi, M., Preneel, B., Sassone, V., Wegener, I. (eds.) ICALP 2006. LNCS, vol. 4052, pp. 1–12. Springer, Heidelberg (2006)
15. Dwork, C.: Differential privacy: a survey of results. In: Agrawal, M., Du, D.-Z., Duan, Z., Li, A. (eds.) TAMC 2008. LNCS, vol. 4978, pp. 1–19. Springer, Heidelberg (2008)
16. Dwork, C.: A firm foundation for private data analysis. Commun. ACM **54**(1), 86–95 (2011)
17. Dwork, C., Rothblum, G.N., Vadhan, S.P.: Boosting and differential privacy. In: 51th Annual IEEE Symposium on Foundations of Computer Science, FOCS 2010, Las Vegas, Neveda, USA, 23–26 October 2010, pp. 51–60 (2010)
18. Dwork, C., Smith, A.: Differential privacy for statistics: What we know and what we want to learn. J. Priv. Confidentiality **1**(2), 135–154 (2009)
19. Haeberlen, A., Pierce, B.C., Narayan, A.: Differential privacy under fire. In: 20th USENIX Security Symposium, San Francisco, CA, USA, 8–12 August 2011, Proceedings, pp. 33–33 (2011)
20. Ho, C.-H., Lin, C.-J.: Large-scale linear support vector regression. J. Mach. Learn. Res. **13**, 3323–3348 (2012)
21. Jordan, M.I.: Divide-and-conquer and statistical inference for big data. In: The 18th ACM SIGKDD International Conference on Knowledge Discovery and Data Mining, KDD 2012, Beijing, China, 12–16 August 2012, p. 4 (2012)
22. Kleiner, A., Talwalkar, A., Sarkar, P., Jordan, M.I.: The big data bootstrap. In: Proceedings of the 29th International Conference on Machine Learning, ICML 2012, Edinburgh, Scotland, UK June 26 - July 1 2012, pp. 1759–1766 (2012)
23. Kraska, T., Talwalkar, A., Duchi, J.C., Griffith, R., Franklin, M.J., Jordan, M.I.: Mlbase: A distributed machine-learning system. In: CIDR 2013, Sixth Biennial Conference on Innovative Data Systems Research, Asilomar, CA, USA, 6–9 January 2013, Online Proceedings (2013)
24. Laptev, N., Zeng, K., Zaniolo, C.: Early accurate results for advanced analytics on mapreduce. Proc. VLDB Endow. PVLDB **5**(10), 1028–1039 (2012)
25. Laptev, N., Zeng, K., Zaniolo, C.: Very fast estimation for result and accuracy of big data analytics: the EARL system. In: 29th IEEE International Conference on Data Engineering, ICDE 2013, Brisbane, Australia, 8–12 April 2013, pp. 1296–1299 (2013)
26. Lin, J., Kolcz, A.: Large-scale machine learning at twitter. In: Proceedings of the ACM SIGMOD International Conference on Management of Data, SIGMOD 2012, AZ, USA,Scottsdale, 20–24 May 2012, pp. 793–804 (2012)
27. Low, Y., Gonzalez, J., Kyrola, A., Bickson, D., Guestrin, C., Hellerstein, J.M.: Distributed graphlab: a framework for machine learning in the cloud. Proc. VLDB Endow. PVLDB **5**(8), 716–727 (2012)
28. McSherry, F.: Privacy integrated queries: an extensible platform for privacy-preserving data analysis. In: Proceedings of the ACM SIGMOD International Conference on Management of Data, SIGMOD 2009, Providence, Rhode Island, USA, June 29 - July 2, 2009, pp. 19–30 (2009)
29. Mohan, P., Thakurta, A., Shi, E., Song, D., Culler, D.E.: Gupt: privacy preserving data analysis made easy. In: Proceedings of the ACM SIGMOD International Conference on Management of Data, SIGMOD 2012, Scottsdale, AZ, USA, 20–24 May 2012, pp. 349–360 (2012)
30. Mukhopadhyay, N.: A consistent and asymptotically efficient two-stage procedure to construct fixed width confidence intervals for the mean. Metrika **27**(1), 281–284 (1980)

31. Mukhopadhyay, Nitis, de Silva, Basil M.: Sequential Methods and Their Applications. Chapman and Hall/CRC, Boca Raton (2008)
32. Nadas, A.: An extension of a theorem of chow and robbins on sequential confidence intervals for the mean. Ann. Math. Stat. **40**(2), 667–671 (1969)
33. Nissim, K., Raskhodnikova, S., Smith, A.: Smooth sensitivity and sampling in private data analysis. In: Proceedings of the 39th Annual ACM Symposium on Theory of Computing, San Diego, California, USA, 11–13 June 2007, pp. 75–84 (2007)
34. Committee on the Analysis of Massive Data: Committee on Applied & Theoretical Statistics, Board on Mathematical Sciences & Their Applications, Division on Engineering & Physical Sciences, and National Research Council. The National Academies Press, Frontiers in Massive Data Analysis (2013)
35. Sandmann, W.: Sequential estimation for prescribed statistical accuracy in stochastic simulation of biological systems. Math. Biosci. **221**(1), 43–53 (2009)
36. Seelbinder, B.M.: On stein's two-stage sampling scheme. Ann. Math. Stat. **24**(4), 640–649 (1953)
37. Smith, A.: Asymptotically optimal and private statistical estimation. In: Garay, J.A., Miyaji, A., Otsuka, A. (eds.) CANS 2009. LNCS, vol. 5888, pp. 53–57. Springer, Heidelberg (2009)
38. Smith, A.: Privacy-preserving statistical estimation with optimal convergence rates. In: Proceedings of the 43rd ACM Symposium on Theory of Computing, STOC 2011, San Jose, CA, USA, June 6–8 2011, pp. 813–822 (2011)
39. Wasserman, L.: Minimaxity, statistical thinking and differential privacy. J. Priv. Confidentiality **4**(1), 51–63 (2012)
40. Xin, R.S., Rosen, J., Zaharia, M., Franklin, M.J., Shenker, S., Stoica, I.: Shark: SQL and rich analytics at scale. In: Proceedings of the ACM SIGMOD International Conference on Management of Data, SIGMOD 2013, New York, NY, USA, 22–27 June 2013, pp. 13–24 (2013)
41. Yui, M., Kojima, I.: A database-hadoop hybrid approach to scalable machine learning. In: IEEE International Congress on Big Data, BigData Congress 2013, June 27 2013-July 2, 2013, pp. 1–8 (2013)
42. Zeng, K., Gao, S., Gu, J., Mozafari, B., Zaniolo, C.: ABS: a system for scalable approximate queries with accuracy guarantees. In: International Conference on Management of Data, SIGMOD 2014, Snowbird, UT, USA, 22–27 June 2014, pp. 1067–1070 (2014)
43. Zeng, K., Gao, S., Mozafari, B., Zaniolo, C.: The analytical bootstrap: a new method for fast error estimation in approximate query processing. In: International Conference on Management of Data, SIGMOD 2014, Snowbird, UT, USA, 22–27 June 2014, pp. 277–288 (2014)

System Modeling and Trust Evaluation
of Distributed Systems

Nagham Alhadad[1], Patricia Serrano-Alvarado[1], Yann Busnel[1,2]([⊠]),
and Philippe Lamarre[3]

[1] LINA/Université de Nantes, Nantes, France
[2] Crest (Ensai) Rennes, Bruz, France
yann.busnel@ensai.fr
[3] LIRIS/INSA Lyon, Écully, France

Abstract. Nowadays, digital systems are connected through complex
architectures. These systems involve persons, physical and digital
resources such that we can consider that a system consists of elements
from two worlds, the social world and the digital world, and their
relations. Users perform activities like chatting, buying, sharing data,
etc. Evaluating and choosing appropriate systems involve aspects like
functionality, performance, QoS, ease of use, or price. Recently, trust
appeared as another key factor for such an evaluation. In this context,
we raise two issues, (i) how to formalize the entities that compose a
system and their relations for a particular activity? and (ii) how to evaluate trust in a system for this activity? This work proposes answers to
both questions. On the one hand, we propose SOCIOPATH, a metamodel
based on first order logic, that allows to model a system considering
entities of the social and digital worlds and their relations. On the other
hand, we propose two approaches to evaluate trust in systems, namely,
SOCIOTRUST and SUBJECTIVETRUST. The former is based on probability theory to evaluate users' trust in systems for a given activity. The
latter is based on subjective logic to take into account uncertainty in
trust values.

1 Introduction

In our daily life, we do social activities like chatting, buying, sending letters,
working, visiting friends, *etc.* These activities are achieved in our society through
physical, digital and human entities. For instance, if we want to *write and send a
letter*, we might type it and print it, we might use a web application to indicate us
the nearest mailbox, then we might consult another web application to provide
us the schedule of the public transport that will allow us to reach the mailbox.

Each entity in this example plays a role enabling us to achieve this activity.
Our PC and printer should enable us to write the letter and print it. The public
transport must allow us to reach the mailbox. The web applications should let
us retrieve the necessary information about our travel. The persons who work
in the postal service should send the letter in a reliable way, *etc.*

© Springer-Verlag Berlin Heidelberg 2015
A. Hameurlain et al. (Eds.): TLDKS XXII, LNCS 9430, pp. 33–74, 2015.
DOI: 10.1007/978-3-662-48567-5_2

Besides the explicit entities we identify, there are implicit ones playing an important role too. For instance, the installed applications on our PC should work properly. Our Internet connection should allow us to access the web applications. The information that we retrieve from the web applications should be reliable. The providers of the physical and digital resources in the postal service should provide reliable resources.

This simple example illustrates that *nowadays activities are achieved thanks to complex systems we* **rely on** *and we* **trust***, maybe unconsciously.*

Many activities are now purely digital (buying online, sharing data, blogging, chatting, and so on). They are supported by systems involving physical and digital resources, *e.g.,* servers, software components, networks, and PCs. However, the fact remains that these resources are provided and controlled by persons (individuals or legal entities) we depend on to execute these activities. The set of these entities and the different relations between them form a complex system for a specific activity. From this point of view, a digital system can be considered as a small society we rely on and we trust to perform our digital activities.

To perform a digital activity, users may face a lot of available options. Many criteria may guide them in their choice: functionality, ease of use, QoS, economical aspects, *etc.* Nowadays, trust is also a momentous aspect of choice.

Starting from these statements, two main issues arise:

1. How to formalize the entities of a system and the relationships between them for a particular activity?
2. How to evaluate trust in a system as a whole for an activity, knowing that a system composes several entities, which can be persons, digital and physical resources?

These points embody the main focus of this study. We argue that studying trust in the separate entities that compose a system does not give a picture of how trustworthy a system is as a whole. The trust in a system depends on its architecture, more precisely, on the way the implicit and explicit entities the users depend on to do their activities, are organized. Thus, the challenge in evaluating trust in a system is firstly, *to model the system architecture for a specific activity.* Secondly, *to define the appropriate metrics to evaluate the user's trust in a modeled system for an activity.*

This paper is organized as follows.[1] In Sect. 2, we propose an answer for the first question with SOCIOPATH, a metamodel that allows to model systems for a digital activity. This metamodel formalizes the entities in a system for an activity and the relations between them. To answer the second question, in Sect. 3, we propose SOCIOTRUST, an approach to evaluate trust in a system for an activity that uses probability theory. And in Sect. 4, we propose a second approach, SUBJECTIVETRUST, where we use subjective logic to take into account uncertainty to evaluate trust. Section 5 presents related works. Finally, we conclude in Sect. 6.

---

[1] Shorter versions of some contributions of this paper have been published in [3–6].

2   SOCIOPATH: Modeling a System

Nowadays, the most widespread architectures belong to the domain of distributed systems. Most of participants' activities on these systems concern their data (sharing and editing documents, publishing photos, purchasing online, *etc.*). As mentioned above, using these systems implies some implicit and explicit relationships, which may be partly unknown. Indeed, users achieve several activities without being aware of the used architecture. In our approach, we believe that users need to have a general representation of the used system including the social and digital entities. Based on this representation, a lot of implicit relations can be deduced like the relations of the social dependence [10,12,29]. With SOCIOPATH [5], we aim to answer the following user's questions about her system:

Q1 Who are the persons that have a possibility to access a user's data? And what are the potential coalitions among persons that could allow undesired access to this data?

Q2 Who are the person(s)/resource(s) a user depends on to perform an activity?

Q3 Who are the persons that can prevent a user from performing an activity?

Q4 Who are the persons that a user is able to avoid to perform an activity?

These questions raise a core last one, *how much a user trusts a system for a specific activity?*

The analysis of systems is usually limited to technical aspects as latency, QoS, functional performance, failure management, *etc.* [9]. The aforementioned questions give some orthogonal but complementary criteria to the classical approach. Currently, people underestimate dependences generated by the systems they use and the resulting potential risks.

Thus, in this section, we propose SOCIOPATH that is based on notions coming from many fields, ranging from computer science to sociology. SOCIOPATH is a generic metamodel that is divided in two worlds: the social world and the digital world. SOCIOPATH allows to draw a representation (or model) of a system that identifies its hardware, software and persons as components, and the ways they are related (*cf.* Sect. 2.1). Enriched with deduction rules (*cf.* Sect. 2.2), SOCIOPATH analyzes the relations between the components and deduces some implicit relations. In SOCIOPATH, we propose some definitions that reveal main aspects about the used architecture for a user (*cf.* Sect. 2.3). An illustrating example shows how SOCIOPATH answers our motivating questions (*cf.* Sect. 2.4).

2.1   SOCIOPATH Metamodel

The SOCIOPATH metamodel allows to describe the architecture of a system in terms of the components that enable people to access digital resources. It distinguishes two worlds; the *social world* and the *digital world*. In the social world, persons or organizations own any kind of physical resources and data. In the digital world, instances of data (including source codes) are stored and processes are running. Figure 1 shows the graphical representation of SOCIOPATH, that we analyze in the next.

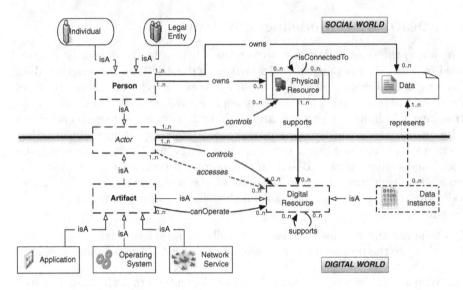

Fig. 1. Graphical view of SOCIOPATH as a UML class diagram.

The Social World includes persons (*e.g.*, users, enterprises, companies), physical resources, data, and relations among them.

- *Data* represent an abstract notion that exists in real life, and does not necessarily imply a physical instance (*e.g.*, address, age, software design).
- *Physical Resource* represents any hardware device (*e.g.*, PC, USB device).
- *Person* represents a generic notion that defines an individual like Alice or a Legal Entity like Microsoft.

The Digital World has entities that are defined as follows:

- *Data Instance* is a digital representation of a *Data* that exists in the social world. For instance, a person has an address (*Data*) in the social world. *Data Instances* of her address can be present in different digital documents: letters ((*e.g.*, encoded using .doc format), contact applications, commercial databases, *etc*. Even if encoded using different formats, each data instance is a semantically equivalent instance of her address. Similarly, a source code is also a *Data Instance* implementing a software (text editor, mailer...) in the digital world.
- *Artifact* represents an abstract notion that describes a "running software". This can be an *Application*, an *Operating System* or a *Network Service*. It may be a single process or a group of processes that should be distributed on different locations, yet defining a single logically coherent entity.
- *Digital Resource* represents an *Artifact* or a *Data Instance*.
- *Actor* represents a *Person* in the social world or an *Artifact* in the digital world. This is the core concept of SOCIOPATH. Indeed, only *Actors* can access or control *Digital Resources* as presented below.

The Relations Proposed in SOCIOPATH are briefly described next. They have to allow to represent in a non naive way how a system is built. They should also help to highlight the links between the structure of a system and the confidence of a user within this system. We do not claim the proposed list to be exhaustive, and one can think about many other relations to describe a system. Providing a complete and minimal set of relations is an interesting question that is out of the scope of this article.

- *owns* is a relation of ownership between a *Person* and a *Physical Resource* (*owns(P, D)*), or between a *Person* and some *Data* (*owns(P,D)*). This relation only exists in the social world.
- *isConnectedTo* is a relation of connection between two *Physical Resources* (*isConnectedTo(PR_1, PR_2)*). It means that two entities are physically connected, through a network for instance. This symmetric relation exists only in the social world.
- *canOperate* represents an *Artifact* that is able to process, communicate or interact correctly with a target *Digital Resource* (*canOperate(F,DR)*). This ability may be explicitly given, for instance, "Microsoft Word" *canOperate* the file letter.doc, or deduced from some general properties, for instance, "Microsoft Word" *canOperate* files of the form *.doc (as far as it can access them - see next relation).
- *accesses* represents an *Actor* that can access a *Digital Resource* (*accesses(A, DR)*). For instance, the operating system accesses the applications installed via this operating system; a person who owns a PC that supports an operating system accesses this operating system. The access relations we consider are: read, write, and execute.
- *controls* represents an *Actor* that can control a *Digital Resource* (*controls(A, DR)*). There should exist different kinds of control relations. For instance, a legal entity, who provides a resource, controls the functionalities of this resource. The persons who use this resource may have some kind of control on it as well. Each of these actors controls the resource in a different way.
- *supports* is a relation between two *Digital Resources* (*supports(DR_1, DR_2)*), or a *Physical Resource* and a *Digital Resource* (*supports(PR, DR)*). It means that the target entity could never exist without the source entity. We may say that the latter allows the former to exist. For instance, an operating system is supported by a given hardware, an application is supported by an operating system, or the code of an application supports this application.
- *represents* is a relation between *Data* in the social world and their *Instances* in the digital world (*represents(D,DI)*). For instance, the source code of the operating system Windows is a representation in the digital world of the data known as "Microsoft Windows" in the social world.

For sake of simplicity, we consider that a person *provides* an artifact, if this person owns the data represented by the data instance which supports the artifact.

Applying SOCIOPATH makes possible non-trivial deductions about relations among entities. For instance, an actor may be able to access digital resources

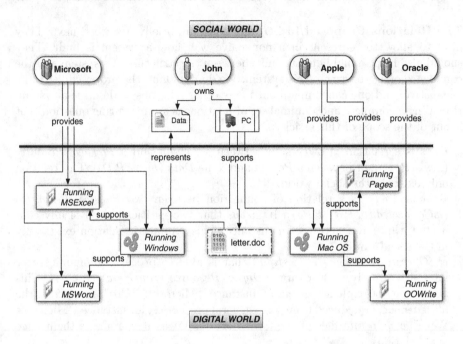

Fig. 2. Use case 1: isolated PC.

supported by different physical resources connected to each other (*e.g.*, a user can access processes running in different hosts).

Use Case 1 of a SocioPath Model: Isolated PC. Fig. 2 shows a simple SocioPath model.[2] In the social world, a user John owns some Data and one PC. There are also legal entities as: Microsoft, provider of Windows, Microsoft Word (MSWord) and Microsoft Excel (MSExcel); Apple, provider of MacOS and Pages; and Oracle, provider of Open Office Writer (OOWrite). In the digital world, two operating systems exist on John's PC: Windows and MacOS. On Windows, two applications are available: MSWord and MSExcel. On MacOS are installed OOWrite and Pages. John's Data are represented in the digital world by the document letter.doc.

We use this example to illustrate some deductions in Sect. 2.2. We deliberately propose a trivial example, in order to show clearly how SocioPath can be applied and how some deductions and definitions are drawn. Table 1 summarizes the notations we use in the following.

2.2 Deduced Access and Control Relations

The semantics of the components and the relations of a SocioPath model allows to deduce more *control* and *access* relations. We use, a first order logic to describe the rules allowing such deductions.

---

[2] In general, we consider that a model conforms to a metamodel.

Table 1. Glossary of notations (1).

Concept	Notation	Set	Remark
Actor	A	$\mathbb{A}$	$A \in \mathbb{A}$
Artifact	F	$\mathbb{F}$	$F \in \mathbb{F}$
Digital resource	DR	$\mathbb{DR}$	$DR \in \mathbb{DR}$
Physical resource	PR	$\mathbb{PR}$	$PR \in \mathbb{PR}$
Data	D	$\mathbb{D}$	$D \in \mathbb{D}$
Data instance	DI	$\mathbb{DI}$	$DI \in \mathbb{DI}$
Operating system	OS	$\mathbb{OS}$	$OS \in \mathbb{OS}$
Path	σ	$\varUpsilon$	$\sigma \in \varUpsilon$
Architecture or system	α	Λ	$\alpha \in \Lambda$
Activity	ω	$\mathbb{W}$	$\omega \in \mathbb{W}$
Activity path	σ^{ω}	$\varUpsilon^{\omega}$	$\sigma^{\omega} \in \varUpsilon^{\omega}$
Activity minimal path	$\widehat{\sigma^{\omega}}$	$\widehat{\varUpsilon^{\omega}}$	$\widehat{\sigma^{\omega}} \in \widehat{\varUpsilon^{\omega}}$
Set of activity restrictions	S	$\mathbb{S}$	$S \in \mathbb{S}$
Person or user	P	$\mathbb{P}$	$P \in \mathbb{P}$

The proposed deduction rules of SOCIOPATH are not exhaustive and by no means we pretend they capture the whole complexity of systems. They capture several aspects of a simplified vision of the systems that serves the purpose of building an understandable and expressive model. Table 2 shows these rules.

– Rule 1 states that if an artifact *can operate* a digital resource and either the artifact and the digital resource are *supported* by the same physical resource or they are *supported* by connected physical resources, then the artifact *accesses* the digital resource.
– Rule 2 states that if a person *owns* a physical resource that *supports* an operating system, then the person *accesses* and *controls* this operating system.
– Rule 3 states that if an operating system *supports* and *can operate* an artifact, then it *controls* this artifact.
– Rule 4 states that if a person *owns* data represented in the digital world by a data instance which *supports* an artifact, then this person *controls* this artifact.
– Rule 5 states the transitivity of relation *accesses*.
– Rule 6 states the transitivity of relation *controls*.
– Rule 7 states that if two physical resources are *connected* to each other, and the first one *supports* an operating system and the second one *supports* another operating system, these two operating systems *access* to each other.

Starting from the use case 1, we apply the SOCIOPATH rules, and obtain the *accesses* and *controls* relations of Fig. 3. For example, from Rule 2, we deduce that John accesses and controls the operating systems MacOS and Windows, and

Table 2. Deduced access and control relations.

Rule	Formal definition
Rule 1	$\forall F \in \mathbb{F}, \forall DR \in \mathbb{DR},$ $\forall PR1, PR2 \in \mathbb{PR}:$ $\bigwedge \begin{cases} canOperate(F, DR) \\ supports(PR1, F) \\ \bigvee \begin{cases} supports(PR1, DR) \\ \bigwedge \begin{cases} supports(PR2, DR) \\ isConnectedTo(PR1, PR2) \end{cases} \end{cases} \end{cases} \Rightarrow accesses(F, DR)$
Rule 2	$\forall P \in \mathbb{P}, \forall PR \in \mathbb{PR}, \forall OS \in \mathbb{OS}: \bigwedge \begin{cases} owns(P, PR) \\ supports(PR, OS) \end{cases} \Rightarrow \bigwedge \begin{cases} accesses(P, OS) \\ controls(P, OS) \end{cases}$
Rule 3	$\forall F \in \mathbb{F}, \forall OS \in \mathbb{OS}: \bigwedge \begin{cases} supports(OS, F) \\ canOperate(OS, F) \end{cases} \Rightarrow controls(OS, F)$
Rule 4·	$\exists P \in \mathbb{P}, \exists D \in \mathbb{D}, \exists DI \in \mathbb{DI}, \exists F \in \mathbb{F}: \bigwedge \begin{cases} owns(P, D) \\ represents(DI, D) \\ supports(DI, F) \end{cases} \Rightarrow controls(P, F)$
Rule 5	$\forall A \in \mathbb{A}, \forall F \in \mathbb{F}, \forall DR \in \mathbb{DR}: \bigwedge \begin{cases} accesses(A, F) \\ accesses(F, DR) \end{cases} \Rightarrow accesses(A, DR)$
Rule 6	$\forall A \in \mathbb{A}, \forall F_1, F_2 \in \mathbb{F}: \bigwedge \begin{cases} controls(A, F_1) \\ controls(F_1, F_2) \end{cases} \Rightarrow controls(A, F_2)$
Rule 7	$\exists PR1, PR2 \in \mathbb{PR},$ $\exists OS1, OS2 \in \mathbb{OS}: \bigwedge \begin{cases} isConnectedTo(PR1, PR2) \\ supports(PR1, OS1) \\ supports(PR2, OS2) \end{cases} \Rightarrow accesses(OS1, OS2)$

from Rule 4, we deduce that Microsoft controls the operating system Windows and Apple controls the operating system MacOS.

2.3   SOCIOPATH Definitions

We next enrich SOCIOPATH with formal definitions to answer the motivating questions (Q1 to Q4) presented in the beginning of this section. Definitions concern activities, paths, and dependences. All of them can be automatically deduced from a SOCIOPATH model.

Definitions for Activities and Paths. A SOCIOPATH model expresses chains of access and control relations, *i.e.,* paths. A user follows a path to perform an activity in a system. In our analysis, we consider systems enabling users to perform a data-based activity. To do so, restrictions must be defined to impose the presence of particular elements in paths. For instance, if a person wants to read a .doc document, she must use an artifact that can "understand" this type of document (*e.g.,* MSWord or OOWrite). Another example, if a person uses a SVN application, the artifacts "SVN client" and "SVN server" should be used and they should appear in the correct order within the path (usually, the SVN client should precede the SVN server).

Definition 1 (Activity ω). *We define an activity ω as a triple $(P, D, \mathcal{S})$, where P is a person, D is a datum and $\mathcal{S}$ is a set of ordered sets $\mathbb{F}$ in a model. So an activity ω is a subset of $\mathbb{P} \times \mathbb{D} \times \mathbb{S}$. The sets in the $\mathbb{S}$ component of an activity are alternative sets of artifacts to perform the activity, i.e., each set allows the person to perform his activity. Thus, $\omega = (P, D, \mathcal{S}) \in \mathbb{P} \times \mathbb{D} \times \mathbb{S}$. For instance, the activity "John edits letter.doc", in use case 1, is defined as ω=(John, Data, {{MSWord}, {Pages}, {OOWrite}}).*

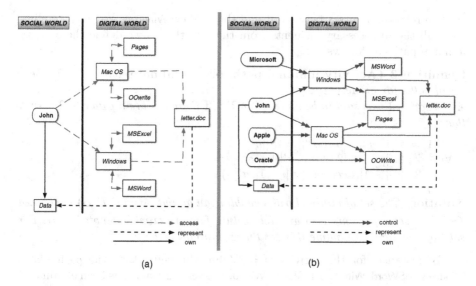

Fig. 3. The relations of *access* and *control* for use case 1 (isolated PC).

We call *paths* the lists of actors and digital resources describing the ways an actor may access a digital resource. A person may perform an activity in different ways and using different intermediate digital resources. Each possibility is described by a path.

Definition 2 (Activity path, or ω-path). *A path σ for an activity $\omega = (P, D, \mathcal{S}) \in \mathbb{P} \times \mathbb{D} \times \mathbb{S}$ is a list of actors and digital resources such that:*

- $\sigma[1] = P$;
- $\sigma[|\sigma|] = D$;
- *represents*$(\sigma[|\sigma| - 1], \sigma[|\sigma|])$;
- $\forall i \in [2 : |\sigma| - 1], (\sigma[i] \in \mathbb{F}) \land accesses(\sigma[i - 1], \sigma[i])$;
- $\exists s \in \mathcal{S}, s \subseteq \sigma$.

Where $\sigma[i]$, denotes the ith element of σ, and $|\sigma|$ the length of σ.

Notation: *Assuming that there is no ambiguity on the model under consideration, the set of ω-paths where $\omega = (P, D, \mathcal{S})$ is denoted Υ^{ω} and the set of all the paths in the model is denoted Υ.*

For example, if John wants to achieve the activity $\omega =$ "John edits letter.doc" using the architecture of use case 1, John uses Windows to work on the application MSWord, which uses Windows file system to access letter.doc so one of the ω-paths for this activity is:

{John, Windows, MSWord, Windows, MSExcel, Windows, letter.doc, Data}.

This path contains some unnecessary artifacts. For instance, MSExcel is an unnecessary artifact to edit letter.doc. It appears in the ω-path because there exists a relation *accesses* between it and the artifact Windows. We want to eliminate all the unnecessary elements from the ω-paths, so we define the activity minimal paths as follows.

Definition 3 (Activity minimal path, or ω-minimal path). *Let Υ^ω be a set of paths for an activity ω.*
A path $\sigma^\omega \in \Upsilon^\omega$ is said to be minimal in Υ^ω iff there exists no path $\sigma' \in \Upsilon^\omega$ such that:

- $\sigma^\omega[1] = \sigma'[1]$ *and* ; $\sigma^\omega[|\sigma^\omega|] = \sigma'[|\sigma'|]$;
- $\forall i \in [2 : |\sigma'|], \exists j \in [2 : |\sigma^\omega|], \sigma'[i] = \sigma^\omega[j]$;
- $\forall i \in [2 : |\sigma'| - 1], accesses(\sigma'[i-1], \sigma'[i])$.

Notation: *The set of minimal paths enabling an activity $\omega = (P, D, \mathcal{S})$ is denoted $\widehat{\Upsilon^\omega}$. This set represents also an architecture for an activity, denoted by α. For sake of simplicity, we name this set the ω-minimal paths.*

For instance, for the activity ω = "John edits letter.doc", the path {John, Windows, MSWord, Windows, MSExcel, Windows, letter.doc, Data} has been eliminated because there is a path σ' = {John, Windows, MSWord, Windows, letter.doc, Data} that satisfies the previous conditions. Thus the set of the ω-minimal paths for this activity are:

$$\alpha = \widehat{\Upsilon^\omega} = \left\{ \begin{array}{l} \{\text{John, Windows, MSWord, Windows, letter.doc, Data}\} \\ \{\text{John, MacOS, OOWrite, Windows, letter.doc, Data}\} \\ \{\text{John, MacOS, Pages, Windows, letter.doc, Data}\} \end{array} \right\}.$$

Definitions for Dependences. Modeling systems with SOCIOPATH allows to underline and discover chains of *accesses* and *controls* relations. In the following, we introduce the definitions of digital dependences (Definitions 4 and 5) and social dependences (Definitions 6 to 9). Informally, the sets of digital dependences of a person are composed of the artifacts a user passes by to reach a particular element. The sets of social dependences are composed of the persons who control these artifacts and the physical resources that support them. We call digital dependences the sets of artifacts a user depends on, because artifacts belong to the digital world in SOCIOPATH. Similarly, we call social dependences the sets of persons and physical resources a user depends on, because they belong to the social world in SOCIOPATH. In the following, these concepts are defined formally and examples refer to use case 1.

Digital Dependences. We say that a person depends on a set of artifacts for an activity ω if each element of this set belongs to one or more paths in the set of the ω-minimal paths.

Definition 4 (Person's dependence on a set of artifacts for an activity). *Let $\omega = (P, D, \mathcal{S})$ be an activity, $\mathcal{F}$ be a set of artifacts and $\widehat{\Upsilon^\omega}$ be the set of ω-minimal paths.*

$$P \text{ depends on } \mathcal{F} \text{ for } \omega \text{ iff } \exists F \subset \mathbb{F}, \forall F \in \mathcal{F}, \exists \sigma \in \widehat{\Upsilon^\omega} : F \in \sigma.$$

For instance, one of the sets John depends on for the activity "John edits letter.doc" is {MacOS, MSWord}.

A person does not depend on all the sets of artifacts in the same way. Some sets may be avoidable because the activity can be executed without them. Some sets are unavoidable because the activity cannot be performed without them. To distinguish the way a person depends on artifacts, we define the degree of a person's dependence on a set of artifacts for an activity as the ratio of the ω-minimal paths that contain these artifacts to all the ω-minimal paths.

Definition 5 (Degree of a person dependence on a set of artifacts).
Let $\omega = (P, D, \mathcal{S})$ be an activity, $\mathcal{F}$ be a set of artifacts and $\widehat{\Upsilon^\omega}$ be the set of ω-minimal paths and $|\widehat{\Upsilon^\omega}|$ is the number of the ω-minimal paths. The degree of dependence of P on $\mathcal{F}$, denoted $d_{\mathcal{F}}^\omega$, is:

$$d_{\mathcal{F}}^\omega = \frac{|\{\sigma : \sigma \in \widehat{\Upsilon^\omega} \wedge \exists F \in \mathcal{F}, F \in \sigma\}|}{|\widehat{\Upsilon^\omega}|}$$

For instance, the degree of dependence of John on the set {MacOS, MSWord} for the activity "John edits letter.doc" is equal to one, while the degree of dependence of John on the set {Pages, OOWrite} is equal to 2/3.

Social Dependences. From the digital dependences, we can deduce the social dependences as follows. A person depends on a set of persons for an activity if the persons in this set control some of the artifacts the person depends on.

Definition 6 (Person's dependence on a set of persons for an activity).
Let $\omega = (P, D, \mathcal{S})$ be an activity, and $\mathcal{P}$ a set of persons.

$$P \text{ depends on } \mathcal{P} \text{ for } \omega \text{ iff } \wedge \begin{cases} \exists \mathcal{F} \subset \mathbb{F} : P \text{ depends on } \mathcal{F} \text{ for } \omega \\ \forall F \in \mathcal{F}, \exists P' \in \mathcal{P} : controls(P', F) \end{cases}$$

For instance, one of the sets of persons John depends on for the activity "John edits letter.doc" is {Oracle, Apple}.

The degree of a person's dependence on a set of persons for an activity is given by the ratio of the ω-minimal paths that contain artifacts controlled by this set of persons.

Definition 7 (Degree of a person's dependence on a set of persons).
Let $\omega = (P, D, \mathcal{S})$ be an activity, $\mathcal{P}$ be a set of persons and $\widehat{\Upsilon^\omega}$ be the ω-minimal paths. The degree of dependence of P on $\mathcal{P}$, denoted $d_{\mathcal{P}}^\omega$, is:

$$d_{\mathcal{P}}^\omega = \frac{|\{\sigma : \sigma \in \widehat{\Upsilon^\omega} \wedge \exists P' \in \mathcal{P}, \exists F \in \sigma, controls(P', F)\}|}{|\widehat{\Upsilon^\omega}|}$$

For instance, the degree of dependence of John on the set {Oracle, Apple} for the activity "John edits letter.doc" is equal to 2/3. We recall that Oracle *controls* OOWrite and Apple *controls* MacOS.

We say a person depends on a set of physical resources for an activity if the elements of this set support the artifacts the person depends on.

Definition 8 (Person's dependence on a set of physical resources). *Let* $\omega = (P, D, \mathcal{S})$ *be an activity, and* $\mathcal{PR}$ *be a set of physical resources.*

$$P \text{ depends on } \mathcal{PR} \text{ for } \omega \text{ iff } \wedge \begin{cases} \exists \mathcal{F} \subset \mathbb{F} : P \text{ depends on } \mathcal{F} \text{ for } \omega \\ \forall F \in \mathcal{F}, \exists PR \in \mathcal{PR} : supports(PR, F) \end{cases}$$

For instance, John depends on the set {PC} for the activity "John edits letter.doc".

The degree of a person's dependence on a set of physical resources for an activity is given by the ratio of the ω-minimal paths that contain artifacts supported by this set of physical resources.

Definition 9 (Degree of a person's dependence on a set of physical resources). *Let* $\omega = (P, D, \mathcal{S})$ *be an activity, let* $\mathcal{PR}$ *be a set of physical resources, let* $\widehat{\varUpsilon^\omega}$ *be the* ω-*minimal paths. The degree of dependence of* P *on* $\mathcal{PR}$, *denoted* $d_{\mathcal{PR}}^\omega$ *is:*

$$d_{\mathcal{PR}}^\omega = \frac{|\{\sigma : \sigma \in \widehat{\varUpsilon^\omega} \wedge \exists PR \in \mathcal{PR}, \exists F \in \sigma, supports(PR, F)\}|}{|\widehat{\varUpsilon^\omega}|}$$

For instance, the degree of dependence of John on the set {PC} for the activity "John edits letter.doc" is equal to 1.

These definitions allow awareness of the user's dependences on the digital and social world. Another use case is presented in the next section to illustrate them.

2.4   Use Case 2 of a SOCIOPATHmodel: GoogleDocs

Figure 4 presents a SOCIOPATH model corresponding to our use case 2 where John uses GoogleDocs for the activity "John reads document.gtxt". In the social world, John owns some Data, a PC and an iPad. We explicitly name only some legal entities who provide resources and artifacts: Microsoft for Windows and Internet Explorer (so called IExplorer), Google for GoogleDocs and Google Cloud services, SkyFireLabs for SkyFire, Apple, for the iOS operating system and the browser Safari and Linux Providers for Linux. NeufTelecom, Orange and SFR are telecom companies. John's iPad is connected to SFR Servers and John's PC is connected to NeufTelecom Servers and Orange Servers. In the digital world, the operating systems Windows and Linux are running on John's PC. Windows supports IExplorer and Linux supports Safari. John's iPad supports the running iOS, which supports two applications, Safari and SkyFire. John's data are represented in the digital world by document.gtxt, which is supported by the physical resources owned by Google. We consider Google Cloud as the storage system used by Google Docs.

Analysis and Results. Through this example we show that SOCIOPATH provides answers to the motivating questions of this section.

Q1 *Who are the persons that have a possibility to access John's data? And what are the potential coalitions among persons that could allow undesired access to this data?*

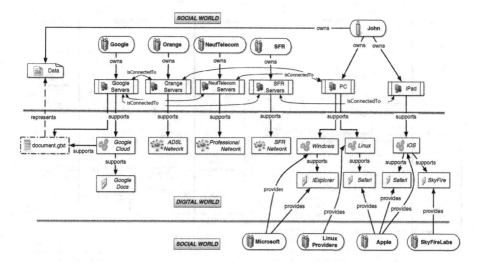

Fig. 4. Use case 2: GoogleDocs.

By applying the deduction rules presented in Sect. 2.3, we deduce the relations of *access* and *control* that exist in this architecture. They are illustrated in Fig. 5. By knowing the relations *accesses* in this model, *cf.* Fig. 5 (a), John is able to know which persons have a possible path to his document. Thus, these persons can[3] access his data. In this example, they are: SFR, NeufTelecom, John, Orange, and Google.

Furthermore, by examining the persons who control the artifacts in the paths, *cf.* Fig. 5 (b), it is possible to understand which coalitions may be done to access John's data. For example, Google can access document.gtxt directly because it controls all the artifacts of the path that enables it to reach it. Orange, instead, has a possible path to access John's data that passes through artifacts controlled by Google. So it must collude with Google to access John's data.

Q2 *Who are the person(s)/resource(s) John depends on to perform the activity "John reads document.gtxt"?*

If John wants to read document.gtxt, he needs a browser and GoogleDocs. So formally, we define this activity as ω=(John, Data, {{SkyFire, GoogleDocs}, {Safari, GoogleDocs}, {IExplorer, GoogleDocs}}). If we apply Definition 3, we find that John has six ω-minimal paths to read document.gtxt:

1. {John, Windows, IExplorer, Windows, ADSL Network, GoogleCloud, GoogleDocs, document.gtxt, Data};
2. {John, Windows, IExplorer, Windows, Professional Network, GoogleCloud, Google-Docs, document.gtxt, Data};

---

[3] By *can*, we mean that a user may be able to perform an action, and not that she has the permissions to do it. In this work, we do not analyze access control and user permission constraints.

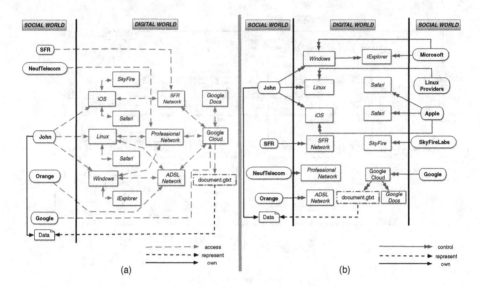

Fig. 5. Relations of access and control in the use case 2 (GoogleDocs).

3. {John, Linux, Safari, Linux, ADSL Network, GoogleCloud, GoogleDocs, document.gtxt, Data};
4. {John, Linux, Safari, Linux, Professional Network, GoogleCloud, GoogleDocs, document.gtxt, Data};
5. {John, iOS, SkyFire, iOS, SFR Network, GoogleCloud, GoogleDocs, document.gtxt, Data};
6. {John, iOS, Safari, iOS, SFR Network, GoogleCloud, GoogleDocs, document.gtxt, Data}.

By applying the definitions of Sect. 2.3, we obtain John's social and digital dependences, and the degree of these dependences for this activity. We show the results concerning some sets of persons John depends on in Table 3 and the degree of dependences on these sets in Fig. 6. This information reveals how much John is autonomous from a specific person or a set of persons. For instance, the degree of dependence on {Microsoft} is 0.33, and the degree of dependence on the set {Apple, NeufTelecom} is 0.83.

Q3 *Who are the persons that can prevent John from performing the activity "John reads document.gtxt"?*
Sets having a degree of dependence equal to 1, are the persons who can prevent John from "reading document.gtxt" because they cross all the ω-paths of this model. These sets are: G8, G9, G10, G12, and G19.

Q4 *Who are the persons that John is able to avoid to perform the activity "John reads document.gtxt"?*
John depends on the sets on which the degree of dependence is less than one, in a less dramatic way (*e.g.*, on the set G8 with a degree of 0.5), because this shows that there are other minimal ω-paths enabling John to read document.gtxt and the persons who belong to this set do not control any artifact

Table 3. Sets of persons John depends on (use case 2 - GoogleDocs).

Group	Sets of persons John depends on	Group	Sets of persons John depends on
G1	{Microsoft}	G12	{Apple,Orange,NeufTelecom}
G2	{Linux Providers}	G13	{Microsoft,SkyFireLabs}
G3	{Apple}	G14	{Orange,SFR}
G4	{SkyFireLabs}	G15	{Apple,Orange}
G5	{SFR}	G16	{Microsoft,NeufTelecom}
G6	{NeufTelecom}	G17	{Microsoft,Orange}
G7	{Orange}	G18	{SkyFireLabs,NeufTelecom}
G8	{Google}	G19	{Microsoft,SFR,Linux Providers}
G9	{Microsoft,Apple}	G20	{Apple,NeufTelecom}
G10	{NeufTelecom,Orange,SFR}	G21	{Linux Providers,SkyFireLabs}
G11	{Linux Providers,SFR}		

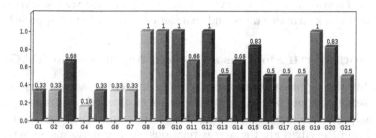

Fig. 6. Degree of dependence on persons' sets.

in these paths. These sets enlighten the "combinations of persons", which John is able to avoid at will.

SOCIOPATH is then useful in the evaluation process of a system with respect to trust requirements. This leads to the fifth question presented in the introduction of this section, namely *How much a user trusts a system for a specific activity?* We focus on answering this question in the following sections.

3   SOCIOTRUST: Evaluating Trust in a System for an Activity Using Probability Theory

Trust has been widely studied in several aspects of daily life. In the trust management community [22,27,30,35–37], two main issues arise, *(i) how to define the trust in an entity, knowing that entities can be persons, digital and physical resources?* and *(ii) how to evaluate such a value of trust in a system under a particular context?* This second point embodies the main focus of this section.

We argue that studying trust in the separate entities that compose a system does not give a picture of how trustworthy a system is as a whole. Indeed, the trust in a system depends on its architecture, more precisely, on the way the implicit and explicit entities, which the users depend on to do their activities, are organized.

Inspired by this idea, we propose SocioTrust [6], an approach to evaluate trust in a system for an activity. The system definition is based on SocioPath models (*cf.* Sect. 2), which here are simplified to present the architecture of a system as a weighted directed acyclic graph (DAG). Levels of trust are then defined for each node in the graph according to the user who evaluates trust. By combining trust values using the theory of probability, we are able to estimate two different granularities of trust, namely, *trust in a path* and *trust in a system*, both for an activity to be performed by a person.

We begin this section introducing how to present a SocioPath model as a directed acyclic graph in Sect. 3.1. Section 3.2 focuses on the main problem that faces trust evaluation that is the existence of *dependent paths*. We propose to solve this problem by using conditional probability. Section 3.4, evaluates our contribution with several experiments that analyze the impact of different characteristics of a system on the behavior of the obtained trust values. Experiments realized on both synthetic traces and real datasets validate our approach.

3.1 A SocioPath Model as a Directed Acyclic Graph (DAG)

We simplify the representation of SocioPath models by aggregating one artifact, the set of persons controlling it, and the set of physical resources supporting it, into only one component. The resulting set of components are the nodes of the DAG and the edges are the *access* relations. A user performs an activity by browsing successive *access* relations through the graph, so-called through *activity minimal paths*.[4]

Definition 10 (A simplified system for an activity, α). *A simplified system that enables a user to achieve an activity, can be expressed as a tuple $\alpha =< N_\omega, A_\omega >$ where:*

- *ω represents the activity the user wants to achieve as a triple (P, D, S) (cf. Sect. 2.3).*
- *N_ω represents the set of nodes n in a system for an activity such that $\{P, D\} \subset N_\omega$, and each triple composed by one artifact, the persons who control it, and the physical resources that support it, are aggregated into one single node, i.e., $n \in N_\omega \setminus \{P, D\}$ such that $n \supseteq \{F, A, PR\}$ iff $controls(A, F) \wedge supports(PR, F)$.*
- *$A_\omega \subseteq N_\omega \times N_\omega$ represents the set of edges in a system. From the rules of SocioPath and the aggregation we made for a node, our DAG exhibits only the relation access.*

---

[4] If there is no ambiguity, we denote an activity minimal path (*i.e.*, ω-minimal path) through the DAG simply by a path σ and each path does not consider the source and the target nodes, *i.e.*, the person and the data instance and the data.

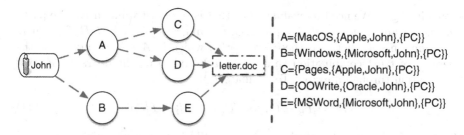

Fig. 7. The system for the activity "John edits letter.doc" as a DAG of use case 1.

Figure 7 illustrates a DAG obtained from the model of the use case 1 for the activity "John edits letter.doc", introduced in Figs. 2 and 3 (*cf.* pages 6 and 9). In this example, all artifacts are supported by the physical resource (PC) owned by John. Here we consider only ω-minimal paths so the path containing the artifact MSExcel is not included. Considered artifacts are Windows, MacOs, MSWord, Pages, and OOWrite. For instance, the node A is a simplification of the artifact MacOS, along with the set of persons who control it {Apple, John} and the set of physical resource that supports it {PC}. Each edge of the DAG represents the relation *accesses*. The paths that enable John to edit letter.doc become: $\sigma_1 = \{A,C\}$; $\sigma_2 = \{A,D\}$; $\sigma_3 = \{B,E\}$. Notice that John, letter.doc, and Data are omitted in this paths' simplification. This type of graph will be used next as well as in Sect. 4.

3.2 The Problem of Dependent Paths

Graph-based trust approaches [1,17,20,21,26,31], are especially used in social networks where the main idea of trust derivation is to propagate trust between two nodes in a graph that represents the social network. In [1], authors propose a general approach where they divide the process of trust evaluation into two steps:

1. Trust combination through a path: the main idea is to combine the trust values among the intermediate edges of a path to obtain a trust value for this path. Several operators are employed ranging from basic operators like the minimum to new operators like *discounting* of subjective logic [19], *cf.* Sect. 4.1.
2. Trust combination through a graph: the main idea is to combine the trust values of all the paths that relate the source with the target, to obtain a single trust value for the graph. Several operators are employed, ranging from basic operators like the average to more recent ones like the *consensus* operator of subjective logic.

In [20,21], Jøsang *et al.* raised a problem of graph-based trust approaches if trust is evaluated through the previous two steps. They argue that some metrics do not give exact results when there are dependent paths, *i.e.*, paths that have common edges in the graph. To explain this problem, we give a simple example shown in Fig. 8. We need to evaluate T_E^A, that is A's trust value in E. The paths between A and E are $path_1 = \{A, B, C, E\}$ and $path_2 = \{A, B, D, E\}$. There is a common edge between these two paths, which is $A \longrightarrow B$. Let $\otimes$ be the operator of trust combination through a path and $\oplus$ be the operator of trust combination through a graph. To evaluate T_E^A:

$$T_E^A = T_B^A \otimes ((T_C^B \otimes T_E^C) \oplus (T_D^B \otimes T_E^D)) \tag{1}$$

However, if we apply the previous two steps, T_E^A is computed as follows:

$$T_E^A = (T_B^A \otimes T_C^B \otimes T_E^C) \oplus (T_B^A \otimes T_D^B \otimes T_E^D) \tag{2}$$

Relations (1) and (2) consist of the same two paths, $path_1$ and $path_2$, but their combined structures are different (T_B^A appears twice in Relation (2)). In some metrics, these two equations produce different results. For instance, when implementing $\otimes$ as binary logic "AND", and $\oplus$ as binary logic "OR", the results would be equal. However, if $\oplus$ is the maximum function and $\otimes$ is the average function, the results are different (*cf.* Fig. 8). It is also the case when $\otimes$ and $\oplus$ are implemented as probabilistic multiplication and comultiplication respectively.

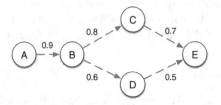

$\otimes$ average $\oplus$ maximum	$\otimes$ multiplication $\oplus$ comultiplication
Relation (1): $T_E^A = 0.825$	Relation (1): $T_E^A = 0.623$
Relation (2): $T_E^A = 0.8$	Relation (2): $T_E^A = 0.64$

Fig. 8. Results of Relations (1) and (2) applied to discrete and continuous metrics.

3.3   A Probabilistic Approach to Infer System Trust Value

If a user needs to evaluate her trust in a system for an activity, she associates each node in the DAG with a trust value and the DAG becomes a weighted directed acyclic graph (WDAG). The notations used here are summarized in Table 4.

We define a function that associates each node with a trust value as $t : \mathbb{N} \rightarrow [0, 1]$ that assigns to each node a person's trust level within the interval $[0, 1]$, where 0 means not trustworthy at all and 1 means fully trustworthy. The values associated to nodes in Fig. 9 are the levels of trust defined by John.

In this study, we adopt the definition of Jøsang *et al.* about trust [22]: *"trust is the probability by which an individual, A, expects that another individual, B, performs a given action on which its welfare depends"*.

According to this, we consider three notions (or granularities) of trust that are formalized in the next.

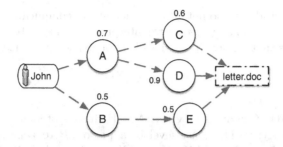

Fig. 9. The activity "John edits letter.doc" as a WDAG.

Table 4. Glossary of notations (2).

Concept	Notation	Remark
Trust in a node for an activity	$t(N)$	For a given
Trust in a path for an activity	$t(\sigma)$	activity ω achieved
Trust in a system for an activity	$t(\alpha)$	by a person P.
Event "N provides the expected services for an activity"	λ^N	The symbols ω and P are omitted in these
Event "P achieves an activity through the path σ"	λ^σ	notations for simplicity.
Event "P achieves an activity through the system"	λ^α	
Probability of an event	$\mathbb{P}(\lambda)$	

- **Trust in a node for an activity**: The trust value of a user P in a node N for an activity ω is the probability, by which P believes that N provides her the expected services for ω. Then, we have $t(N) = \mathbb{P}(\lambda^N)$.
- **Trust in a path for an activity**: The trust value of a user P in a path σ for an activity ω is the probability, by which P believes that σ enables her to achieve ω. Then, we have $t(\sigma) = \mathbb{P}(\lambda^\sigma)$.
- **Trust in a system for an activity**: The trust value of a user P in a system α for an activity ω is the probability, by which P believes that α enables her to achieve ω. Then, we have $t(\alpha) = \mathbb{P}(\lambda^\alpha)$.

Trust in a Node for an Activity. Trust in a node is evaluated from the point of view of the concerned user. There are several ways to construct this trust level. We can figure out different objective and subjective factors that impact this trust level, like the reputation of the persons who control the artifact, their skills, the performance of the physical resource that supports the artifact or the personal experience with this artifact. We thus have $t(N) = f(t_\omega^F, t_\omega^P, t_\omega^{PR})$, where t_ω^F, t_ω^P, t_ω^{PR} are respectively the trust values assigned to an artifact F, the set of persons $\mathcal{P}$ who control F, and the set of physical resources $\mathcal{PR}$ that supports F for a given activity ω. The meaning of the resulting trust value in a node depends on the employed function f to compute this value [28]. For instance, if Bayesian inference is employed to evaluate it as is done in [24], the node trust value is considered as *"the probability by which a user believes that a node can perform an expected action for a given activity"* [13].

However, in this study, we do not address the issue of computing the trust value of a user in a node for an activity but we interpret it as the probability, by which a user P believes that a node N provides her the expected services for ω. Then, we have:

$$t(N) = \mathbb{P}(\lambda^N) \qquad (3)$$

Trust in a Path for an Activity. A path in a system represents a way to achieve an activity. The trust level of a person P to achieve an activity through a particular path $\sigma = \{N_1, N_2, \ldots, N_n\}$ is the probability that all nodes $\{N_i\}_{i \in [1..n]}$ provide the expected services for the activity. Thus $\mathbb{P}(\lambda^\sigma)$ is computed as follows:

$$t(\sigma) = \mathbb{P}(\lambda^\sigma) = \mathbb{P}(\lambda^{N_1} \wedge \lambda^{N_2} \wedge \ldots \wedge \lambda^{N_n})$$

The event λ^{N_i} means that N_i provides the expected services for an activity. Since the graph is acyclic (only minimum activity paths are considered), then the nodes $N_1, \ldots, N_n$ are different in the path, thus each λ^{N_i} is independent from all others. Hence, we can rewrite the trust in a path as follows:

$$t(\sigma) = \mathbb{P}(\lambda^\sigma) = \mathbb{P}(\lambda^{N_1}) \times \mathbb{P}(\lambda^{N_2}) \times \ldots \times \mathbb{P}(\lambda^{N_n}) = \prod_{i=1}^{n} \mathbb{P}(\lambda^{N_i}) \qquad (4)$$

Trust in a System for an Activity. In general, a system is composed of several paths that represent the different ways a person has, to achieve an activity. The trust level of a person P in a system α to achieve an activity is the probability that she achieves her activity through at least one of the paths in the system. To evaluate the trust in a system for an activity, two cases have to be considered: (i) the paths are independent, i.e., they do not have nodes in common[5] and (ii) the paths are dependent, i.e., paths having nodes in common.

Independent Paths. Let $\{\sigma_i\}_{i \in [1..m]}$ be independent paths that enable a person P to achieve an activity. The probability of achieving the activity through a system, $\mathbb{P}(\lambda^\alpha)$, is the probability of achieving the activity through at least one of the paths σ_i. Thus $\mathbb{P}(\lambda^\alpha)$ is computed as follows:

$$t(\alpha) = \mathbb{P}(\lambda^\alpha) = \mathbb{P}(\lambda^{\sigma_1} \vee \lambda^{\sigma_2} \vee \ldots \vee \lambda^{\sigma_m})$$

Since the paths are independent then the equation can be rewritten as follows:

$$t(\alpha) = \mathbb{P}(\lambda^\alpha) = 1 - \prod_{i=1}^{m} (1 - \mathbb{P}(\lambda^{\sigma_i})) \qquad (5)$$

---

[5] The dependent paths in our graph are the paths that have common nodes (and not common edges) because the trust value is associated to a node, and not to an edge as in a social network.

For instance, if a person has two independent paths to achieve an activity then:

$$t(\alpha) = \mathbb{P}(\lambda^\alpha) = \mathbb{P}(\lambda^{\sigma_1} \vee \lambda^{\sigma_2})$$
$$= 1 - (1 - \mathbb{P}(\lambda^{\sigma_1})) \times (1 - \mathbb{P}(\lambda^{\sigma_2})) \tag{6}$$
$$= \mathbb{P}(\lambda^{\sigma_1}) + \mathbb{P}(\lambda^{\sigma_2}) - \mathbb{P}(\lambda^{\sigma_1}) \times \mathbb{P}(\lambda^{\sigma_2})$$

Dependent Paths. When there are common nodes between paths, Relation (5) cannot be applied directly. To evaluate the trust through dependent paths, we begin with a simple case, where a system has two paths, before generalizing.

1. **Two dependent paths with one common node.** Let σ_1, σ_2, be two paths that enable a person P to achieve an activity. $\sigma_1 = \{N, N_{1,2}, \ldots, N_{1,n}\}$, $\sigma_2 = \{N, N_{2,2}, \ldots, N_{2,m}\}$. These two paths have a common node, which is N so they are dependent. Thus the probability that a person P achieves the activity ω through the system α is computed as follows:

$$t(\alpha) = \mathbb{P}(\lambda^\alpha) = \mathbb{P}(\lambda^{\sigma_1} \vee \lambda^{\sigma_2}) = \mathbb{P}(\lambda^{\sigma_1}) + \mathbb{P}(\lambda^{\sigma_2}) - \mathbb{P}(\lambda^{\sigma_1} \wedge \lambda^{\sigma_2})$$

The probability $\mathbb{P}(\lambda^{\sigma_1} \wedge \lambda^{\sigma_2})$ can be rewritten using conditional probability as the two paths are dependent.

$$t(\alpha) = \mathbb{P}(\lambda^\alpha) = \mathbb{P}(\lambda^{\sigma_1}) + \mathbb{P}(\lambda^{\sigma_2}) - \mathbb{P}(\lambda^{\sigma_2}) \times \mathbb{P}(\lambda^{\sigma_1}|\lambda^{\sigma_2})$$
$$= \mathbb{P}(\lambda^{\sigma_1}) + \mathbb{P}(\lambda^{\sigma_2}) \times (1 - \mathbb{P}(\lambda^{\sigma_1}|\lambda^{\sigma_2}))$$

We have to compute $\mathbb{P}(\lambda^{\sigma_1}|\lambda^{\sigma_2})$, which is the probability that P achieves the activity through σ_1 once it is already known that P achieves the activity through σ_2. Thus, it is the probability that N, $\{N_{1,i}\}_{i\in[2..n]}$ provides the expected services for this activity, once it is known that N, $\{N_{2,i}\}_{i\in[2..m]}$ provided the expected services. Thus, N has already provided the expected services. Hence, $\mathbb{P}(\lambda^{\sigma_1}|\lambda^{\sigma_2}) = \prod_{i=2}^{n} \mathbb{P}(\lambda^{N_{1,i}})$, where $\lambda^{N_{1,i}}$ is the event "$N_{1,i}$ provides the necessary services for the activity".

$$t(\alpha) = \mathbb{P}(\lambda^\alpha)$$

$$= \mathbb{P}(\lambda^N) \times \prod_{i=2}^{n} \mathbb{P}(\lambda^{N_{1,i}}) + \mathbb{P}(\lambda^N) \times \prod_{i=2}^{m} \mathbb{P}(\lambda^{N_{2,i}}) \times (1 - \prod_{i=2}^{n} \mathbb{P}(\lambda^{N_{1,i}}))$$

$$= \mathbb{P}(\lambda^N) \times \left[\prod_{i=2}^{n} \mathbb{P}(\lambda^{N_{1,i}}) + \prod_{i=2}^{m} \mathbb{P}(\lambda^{N_{2,i}}) \times (1 - \prod_{i=2}^{n} \mathbb{P}(\lambda^{N_{1,i}})) \right]$$

$$= \mathbb{P}(\lambda^N) \times \left[\prod_{i=2}^{n} \mathbb{P}(\lambda^{N_{1,i}}) + \prod_{i=2}^{m} \mathbb{P}(\lambda^{N_{2,i}}) - \prod_{i=2}^{m} \mathbb{P}(\lambda^{N_{2,i}}) \times \prod_{i=2}^{n} \mathbb{P}(\lambda^{N_{1,i}}) \right]$$

From Relation (6) we can note that the term:

$$\prod_{i=2}^{n} \mathbb{P}(\lambda^{N_{1,i}}) + \prod_{i=2}^{m} \mathbb{P}(\lambda^{N_{2,i}}) - \prod_{i=2}^{m} \mathbb{P}(\lambda^{N_{2,i}}) \times \prod_{i=2}^{n} \mathbb{P}(\lambda^{N_{1,i}})$$

is the probability that P achieves the activity through $\sigma_1' = \{N_{1,2}, \ldots, N_{1,n}\}$ or $\sigma_2' = \{N_{2,2}, \ldots, N_{2,m}\}$, which are the paths after eliminating the common nodes. Thus the previous equation can be rewritten as follows:

$$t(\alpha) = \mathbb{P}(\lambda^\alpha) = \mathbb{P}(\lambda^N) \times \mathbb{P}(\lambda^{\sigma_1'} \vee \lambda^{\sigma_2'})$$

2. **Two dependent paths with several common nodes.** Let σ_1, σ_2, be two paths that enable a person P to achieve an activity. These two paths have several common nodes. By following the same logic as before, we compute the probability that a person P achieves activity ω through system α as follows:

$$t(\alpha) = \mathbb{P}(\lambda^\alpha) = \prod_{N \in \sigma_1 \cap \sigma_2} \mathbb{P}(\lambda^N) \times \mathbb{P}(\lambda^{\sigma_1'} \vee \lambda^{\sigma_2'})$$

where $\sigma_1' = \sigma_1 \setminus \sigma_2$, $\sigma_2' = \sigma_2 \setminus \sigma_1$.

3. **Several dependent paths.** A person may have several paths l with common nodes. Thus $\mathbb{P}(\lambda^\alpha)$ is computed as follows:

$$t(\alpha) = \mathbb{P}(\lambda^\alpha) = \mathbb{P}(\lambda^{\sigma_1} \vee \lambda^{\sigma_2} \vee \ldots \vee \lambda^{\sigma_l}) =$$

$$\mathbb{P}(\lambda^{\sigma_1} \vee \lambda^{\sigma_2} \vee \ldots \vee \lambda^{\sigma_{l-1}}) + \mathbb{P}(\lambda^{\sigma_l}) - \mathbb{P}(\lambda^{\sigma_l}) \times \mathbb{P}(\lambda^{\sigma_1} \vee \lambda^{\sigma_2} \vee \ldots \vee \lambda^{\sigma_{l-1}} | \lambda^{\sigma_l}) \quad (7)$$

Let us discuss these terms one by one:

- The term $\mathbb{P}(\lambda^{\sigma_l})$ can be computed directly from Relation (4).
- The term $\mathbb{P}(\lambda^{\sigma_1} \vee \lambda^{\sigma_2} \vee \ldots \vee \lambda^{\sigma_{l-1}})$ can be computed recursively using Relation (7).
- The term $\mathbb{P}(\lambda^{\sigma_1} \vee \lambda^{\sigma_2} \vee \ldots \vee \lambda^{\sigma_{l-1}} | \lambda^{\sigma_l})$ needs first to be simplified. If we follow the same logic as before, the term $\mathbb{P}(\lambda^{\sigma_1} \vee \lambda^{\sigma_2} \vee \ldots \vee \lambda^{\sigma_{l-1}} | \lambda^{\sigma_l})$ can be replaced by the term $\mathbb{P}(\lambda^{\sigma_1'} \vee \lambda^{\sigma_2'} \vee \ldots \vee \lambda^{\sigma_{l-1}'})$ where we obtain each $\lambda^{\sigma_i'}$ by eliminating the nodes in common with σ_l.
- $\mathbb{P}(\lambda^{\sigma_1'} \vee \lambda^{\sigma_2'} \vee \ldots \vee \lambda^{\sigma_{l-1}'})$ can be computed recursively using Relation (7), and recursion is guaranteed to terminate while the number of paths is finite.

We are now able to evaluate the trust in a whole system α.

3.4   Experimental Evaluations

In this section, we present different experiments, their results, analysis, and interpretation. The main objectives are *(i)* to study the influence of the system organization on the computed trust values and *(ii)* to confront this approach with real users.

Influence of the System Architecture on the Trust Value. This experiment studies the influence of the system organization on the computed trust value. We apply our equations on different systems that have the same number of nodes and the same values of trust assigned to each node, but assembled in different topologies as presented in Table 5. The values of trust associated to nodes A,B,C,D,E,F are $0.1, 0.2, 0.3, 0.9, 0.8, 0.7$ respectively.

Table 5. Different systems and their trust values.

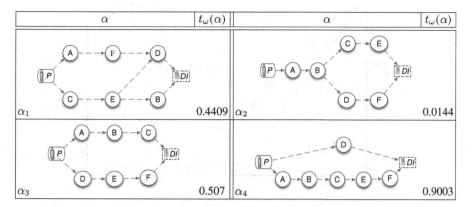

α	$t_\omega(\alpha)$	α	$t_\omega(\alpha)$
α_1	0.4409	α_2	0.0144
α_3	0.507	α_4	0.9003

We compute the trust value $t(\alpha)$ for each system. We obtain very divergent results varying from 0.0144 to 0.9003 as illustrated in Table 5. Thus, collecting the values of trust in each separated node in a system is not enough to determine if the system is trustworthy or not for an activity. One must also know how the system is organized. For example, in α_2, all the paths contain the nodes A and B and the trust values in these nodes are quite low, 0.1 and 0.2 respectively, so the system trust value is also low due to the strong dependency on these two nodes in this system.

Influence of the Path Length and the Number of Paths on the Trust Value. This experiment observes the evolution of the trust value for an activity according to some characteristics of the graph like path's length and number of paths. As a dataset, we consider random graphs composed of 20 to 100 nodes, and 1 to 15 paths. Each node in the graph is associated to a random value of trust from a predefined range.

First, the evolution of trust values according to the paths' lengths in a graph is evaluated. Each simulated graph is composed of 5 paths with lengths varying from 1 to 15 nodes. Different trust values were simulated in the ranges $[0.6, 0.9]$, $[0.1, 0.9]$ *etc.* Figure 10 illustrates the impact of the path length on the trust value. Note that, the system trust value decreases when the length of paths increases. This reflects a natural intuition we had from the fact that trust values are multiplied.

Second, we set the path lengths to 5 nodes and we increased the number of paths from 1 up to 15 in order to observe the variation of the trust values. Again, different node trust values were simulated in the ranges $[0.7, 0.9]$, $[0.6, 0.9]$, *etc.* Figure 11 illustrates that the trust value increases as the number of paths increases. This reflects the intuition that the measure of trust in a system for an activity rises when the number of ways to achieve this activity increases.

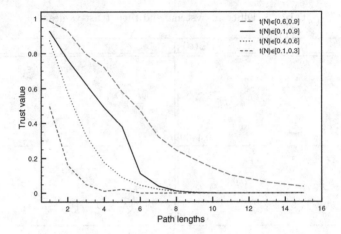

Fig. 10. System trust value according to the length of paths.

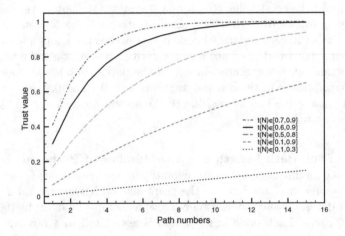

Fig. 11. System trust value according to the number of paths.

Social Evaluation (a Real Case). In order to evaluate our proposal in a real use case, we modeled part of the SVN system of LINA research laboratory[6] with SOCIOPATH. SVN (Subversion) is a client-server system to manage versions of files. The server allocates repositories of files and clients make copies of repositories. Copies of files contained in repositories can be modified at the client side, modification must be committed to generate new versions. Other clients must frequently update their copies. Persons on which SVN users depend on, are the LINA laboratory that owns the server and the software SVN, the engineer that controls the software at the server side of the SVN, the provider of the software SVN, the computer and the software at the client side, *etc.* We applied

---

[6] https://www.lina.univ-nantes.fr/.

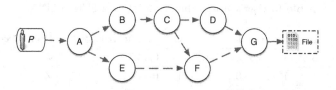

Fig. 12. LINA's WDAG for the activity "a user accesses a file on the SVN".

the rules of SOCIOPATH on this system for the activity "a user accesses a file on the SVN". Due to privacy issues, Fig. 12 presents the DAG for this activity with anonymous nodes. For the sake of clarity, we simplify the underlying graph as much as possible.

Based on this context, we conducted an opinion survey among twenty members of LINA including, PhD students, professors and technicians about their level of trust in each node. For each person, we have computed the system trust value according to the methodology presented in Sect. 3.3. Table 6 presents the data of the survey and the computed trust values. In a second phase, we asked each user if the SOCIOTRUSTproposal correctly reflects her trust towards the SVN system used in our lab. The possibilities of answer were simply Yes or No. The last column of Table 6 presents this feedback, where ✓ means that they are satisfied, and × means that they are not satisfied. 75 % of the users are satisfied with the computation. Unsatisfied users argue that they expected a higher trust value. Some of the trust values associated to the nodes of the unsatisfied users, have relatively low values (around 0.5 or 0.6) compared to other users. These users explained that the lack of knowledge about some nodes leads them to define what they called *a neutral value* (*i.e.,* 0.5 or 0.6) that they considered neither trustworthy, nor untrustworthy. Clearly, such a behavior is not compatible with a probabilistic interpretation where 0.5 is like any other possible value between 0 and 1 and has nothing of neutral.

The explanations provided by users revealed an interesting point: even in a small environment and considering advanced users, no one is in possession of all the information necessary to construct an informed assessment. To conform to this reality and model this phenomenon, it is necesary to use a formalism allowing to express uncertainty related to incompleteness of available information. Extending our approach to use subjective logic [19], which can express uncertainty or ignorance, is the objective of the next section.

4   SUBJECTIVETRUST: Evaluating Trust in a System for an Activity Using Subjective Logic

SOCIOTRUST is oriented to full-knowledge environments. However, in uncertain environments, users might not be in possession of all the information to provide a dogmatic opinion and traditional probability cannot express uncertainty. With subjective logic [19], trust can be expressed as subjective opinions with degrees of

Table 6. User's trust value in the system SVN in LINA.

	A	B	C	D	E	F	G	System trust value	User's feedback
P_1	0.5	0.5	1	0.5	0.5	1	1	0.4375	✓
P_2	0.7	1	1	0.7	0.7	1	1	0.847	✓
P_3	0.5	0.5	1	0.7	0.5	1	1	0.4375	✗
P_4	0.6	0.6	0.8	0.7	0.6	0.8	0.6	0.3072	✗
P_5	0.8	0.8	1	0.8	0.8	1	0.9	0.8202	✓
P_6	0.9	0.9	1	0.9	0.9	0.9	0.9	0.9043	✓
P_7	0.6	0.6	0.7	0.6	0.6	0.6	0.7	0.2770	✗
P_8	0.8	0.6	1	0.9	0.8	0.8	1	0.7416	✓
P_9	0.7	0.5	1	0.4	0.7	0.6	0.9	0.4407	✓
P_{10}	0.8	1	0.7	0.8	0.8	0.9	0.8	0.6975	✓
P_{11}	0.5	0.5	0.9	0.5	0.5	0.5	0.9	0.2473	✗
P_{12}	0.95	0.95	0.8	0.8	0.95	0.95	0.8	0.8655	✓
P_{13}	0.8	0.9	0.8	0.7	0.95	0.8	0.7	0.6433	✓
P_{14}	0.8	0.7	0.9	0.7	0.9	0.8	0.8	0.6652	✓
P_{15}	0.9	0.8	0.8	0.9	0.9	0.9	0.8	0.7733	✓
P_{16}	0.7	0.6	0.6	0.6	0.8	0.7	0.6	0.337	✓
P_{17}	0.5	0.9	0.8	0.7	0.9	0.5	0.8	0.3807	✗
P_{18}	0.7	0.7	1	0.7	0.6	0.7	1	0.6088	✓
P_{19}	0.8	0.8	1	1	1	0.8	1	0.8704	✓
P_{20}	0.9	0.9	0.8	0.9	0.9	0.9	0.8	0.7971	✓

uncertainty. In this section, we aim to take advantage of the benefits of subjective logic to evaluate trust.

The main contribution of this section is proposing a generic model named SUBJECTIVETRUST [4], for evaluating trust in a system for an activity taking into account uncertainty. By combining the user's opinion on a node, we are able to estimate two different granularities of trust, namely, *opinion on a path* and *opinion on a system*, both for an activity to be performed by a person. As we know, the main problem that faces trust evaluation based on a graph is the existence of *dependent paths*. To solve this problem, we propose two methods: **Copy** and **Split**.

Next section presents some preliminaries about subjective logic, then we present SUBJECTIVETRUST in Sect. 4.2 and finally some experiments in Sect. 4.3.

4.1 Preliminaries About Subjective Logic

In the terminology of subjective logic [19], an opinion held by an individual P about a proposition x is the ordered quadruple $O_x = (b_x, d_x, u_x, a_x)$ where:

- b_x (belief) is the belief that x is true.
- d_x (disbelief) is the belief that the x is false.
- u_x (uncertainty) is the amount of uncommitted belief.
- a_x is called the base rate, it is the a priori probability in the absence of evidence.

Note that $b_x, d_x, u_x, a_x \in [0, 1]$ and $b_x + d_x + u_x = 1$. a_x is used for computing an opinion's probability expectation value that can be determined as $\mathbb{E}(O_x) = b_x + a_x u_x$. More precisely, a_x determines how uncertainty shall contribute to the probability expectation value $\mathbb{E}(O_x)$.

Subjective logic consists of a set of logical operations which are defined to combine opinions.

- Conjunction operator ($\wedge$) represents the opinion of a person on several propositions.
- Disjunction operator ($\vee$) represents the opinion of a person on one of the propositions or any union of them.
- Discounting operator ($\otimes$) represents the transitivity of the opinions.
- Consensus operator ($\oplus$) represents the consensus of opinions of different persons.

In our work, we rely on a graph to evaluate trust like in the social network domain, but our interpretation of the graph is different. For us, a graph represents *a system for a digital activity* and not *a social network*. This assumption plays an important role in the operations we apply for trust evaluation. That is why, in a social network, to evaluate trust through a path using subjective logic, the operator of discounting ($\otimes$) is used to compute the transitivity through a path, whereas, in our work, evaluating trust in a path is the trust in the *collection* of the nodes that form this path, *i.e.*, conjunction. In the same manner, to evaluate trust through a graph in a social network, the operator of consensus ($\oplus$) is used to evaluate the consensus of opinions of different persons through the different paths that form the graph, whereas, in our work, paths represent the ways one person disposes to achieve an activity, so evaluating trust in a graph is the trust in at least one of the paths or any union of them, *i.e.*, disjunction. In the following, we present the conjunction and disjunction operators that we use in SUBJECTIVETRUST.

- Conjunction represents the opinion of a person on several propositions. Let $O_x^P = (b_x^P, d_x^P, u_x^P, a_x^P)$ and $O_y^P = (b_y^P, d_y^P, u_y^P, a_y^P)$ be respectively P's opinion on x and y. $O_{x \wedge y}^P$ represents P's opinion on both x and y and can be calculated as follows:

$$O_x^P \wedge O_y^P = O_{x \wedge y}^P = \begin{cases} b_{x \wedge y}^P = b_x^P b_y^P \\ d_{x \wedge y}^P = d_x^P + d_y^P - d_x^P d_y^P \\ u_{x \wedge y}^P = b_x^P u_y^P + u_x^P b_y^P + u_x^P u_y^P \\ a_{x \wedge y}^P = \frac{b_x^P u_y^P a_y^P + b_y^P u_x^P a_x^P + u_x^P a_x^P u_y^P a_y^P}{b_x^P u_y^P + u_x^P b_y^P + u_x^P u_y^P} \end{cases} \quad (8)$$

$$\mathbb{E}(O_x^P \wedge O_y^P) = \mathbb{E}(O_{x \wedge y}^P) = \mathbb{E}(O_x^P)\mathbb{E}(O_y^P) \quad (9)$$

– Disjunction represents the opinion of a person on one of the propositions or any union of them. Let $O_x^P = (b_x^P, d_x^P, u_x^P, a_x^P)$ and $O_y^P = (b_y^P, d_y^P, u_y^P, a_y^P)$ be respectively P's opinion on x and y. $O_{x\lor y}^P$ represents P's opinion on x or y or both and can be calculated with the following relations:

$$O_x^P \lor O_y^P = O_{x\lor y}^P = \begin{cases} b_{x\lor y}^P = b_x^P + b_y^P - b_x^P b_y^P \\ d_{x\lor y}^P = d_x^P d_y^P \\ u_{x\lor y}^P = d_x^P u_y^P + u_x^P d_y^P + u_x^P u_y^P \\ a_{x\lor y}^P = \frac{u_x^P a_x^P + u_y^P a_y^P - b_x^P u_y^P a_y^P - b_y^P u_x^P a_x^P - u_x^P a_x^P u_y^P a_y^P}{u_x^P + u_y^P - b_x^P u_y^P - b_y^P u_x^P - u_x^P u_y^P} \end{cases} \tag{10}$$

$$\mathbb{E}(O_x^P \lor O_y^P) = \mathbb{E}(O_{x\lor y}^P) = \mathbb{E}(O_x^P) + \mathbb{E}(O_y^P) - \mathbb{E}(O_x^P)\mathbb{E}(O_y^P) \tag{11}$$

It is important to mention that conjunction and disjunction are commutative and associative.

$$O_x^P \land O_y^P = O_y^P \land O_x^P$$

$$O_x^P \lor O_y^P = O_y^P \lor O_x^P$$

$$(O_x^P \land O_y^P) \land O_z^P = O_x^P \land (O_y^P \land O_z^P)$$

$$(O_x^P \lor O_y^P) \lor O_z^P = O_x^P \lor (O_y^P \lor O_z^P)$$

However, the conjunction over the disjunction is not distributive. This is due to the fact that opinions must be assumed to be independent, whereas distribution always introduces an element of dependence.

$$O_x^P \land (O_y^P \lor O_z^P) \neq (O_x^P \land O_y^P) \lor (O_x^P \land O_z^P)$$

By using these operators, in the next section we combine the opinions on the nodes to estimate two different granularities of trust: opinion on a path and opinion on a system.

4.2   Inferring User's Opinion on a System Using Subjective Logic

This section presents SUBJECTIVETRUST, a graph-based trust approach to infer trust in a system for an activity using subjective logic.

The system definition is based on SOCIOPATH. To focus on trust in the system, the SOCIOPATH model is abstracted in a DAG as in SOCIOTRUST (*cf.* Sect. 3.1). In subjective logic, trust is expressed as an opinion, thus in this proposition, the DAG is weighted with opinions, *i.e.*, each node is associated with an opinion in the form (b, d, u, a). Figure 13 shows the WDAG of use case 1, where the values associated to nodes represent John's opinion on these nodes.

As in SOCIOTRUST, opinion on a node is evaluated from the point of view of the concerned user depending on her personal experience with this node. Several approaches have been proposed to obtain this opinion. In [19], authors translate the user's negative or positive observations to opinions. In [24,25], the opinion parameters are estimated by Bayesian inference. In this study, we do not address

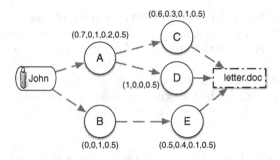

Fig. 13. The activity "John edits letter.doc" as a WDAG where weights are opinions (use case 1).

the issue of obtaining this opinion, we focus on combining the opinions associated on the nodes to obtain an opinion on a path and on a system for an activity.

Next sections show how an opinion on a path and an opinion on a system are evaluated by combining respectively the opinions on the nodes and the opinions on the paths, using the appropriate operators of subjective logic.

Opinion on a Path for an Activity. A path in a system represents a way to achieve an activity. An opinion on a path that contains several nodes can be computed by combining the opinions on the nodes that belong to it.

In trust propagation, the operator to build an opinion on a path is discounting because it allows to compute the transitivity of an opinion along a path [20, 21]. However, if a person needs to achieve an activity through a path, she needs to pass through all the nodes composing this path. Hence, an opinion on a path is the opinion on all nodes composing this path.

The conjunction operator represents the opinion of a person on several propositions. Thus, it is appropriate to compute an opinion on a path from the opinions on the nodes.

Let $\sigma = \{N_1, N_2, \ldots, N_n\}$ be a path that enables a user P to achieve an activity. P's opinion on the nodes $\{N_i\}_{i \in [1..n]}$ for an activity are denoted by $O_{N_i} = (b_{N_i}, d_{N_i}, u_{N_i}, a_{N_i})$. P's opinion on the path σ for achieving an activity, denoted by $O_\sigma = (b_\sigma, d_\sigma, u_\sigma, a_\sigma)$, can be derived by the conjunction of P's opinions on $\{N_i\}_{i \in [1..n]}$. $O_{\sigma = \{N_1, \ldots, N_n\}} = \bigwedge \{O_{N_i}\}_{i \in [1..n]}$. Given Relation (8), we obtain the following generalization: $O_{\sigma = \{N_1, \ldots, N_n\}} =$

$$
\begin{cases}
b_{\sigma = \{N_1, \ldots, N_n\}} = b_{\bigwedge \{N_i\}_{i \in [1..n]}} = \prod_{i=1}^{n} b_{N_i} \\
d_{\sigma = \{N_1, \ldots, N_n\}} = d_{\bigwedge \{N_i\}_{i \in [1..n]}} = 1 - \prod_{i=1}^{n} (1 - d_{N_i}) \\
u_{\sigma = \{N_1, \ldots, N_n\}} = u_{\bigwedge \{N_i\}_{i \in [1..n]}} = \prod_{i=1}^{n} (b_{N_i} + u_{N_i}) - \prod_{i=1}^{n} (b_{N_i}) \\
a_{\sigma = \{N_1, \ldots, N_n\}} = a_{\bigwedge \{N_i\}_{i \in [1..n]}} = \frac{\prod_{i=1}^{n} (b_{N_i} + u_{N_i} a_{N_i}) - \prod_{i=1}^{n} (b_{N_i})}{\prod_{i=1}^{n} (b_{N_i} + u_{N_i}) - \prod_{i=1}^{n} (b_{N_i})}
\end{cases}
\tag{12}
$$

Due to space constraint and as they are almost straightforward, the proofs of Relation (12) and the verifications of the correction (*i.e.*, $b_\sigma + d_\sigma + u_\sigma = 1$, $0 < b_\sigma < 1$, $0 < d_\sigma < 1$, $0 < u_\sigma < 1$ and $0 < a_\sigma < 1$) are presented in [3].

Opinion on a System for an Activity. In trust propagation, to build an opinion on a target node in a graph, the consensus operator is used because it represents the consensus of the opinions of different persons through different paths [20,21]. In our work, an opinion on a system is the opinion of a person on one or several paths. Thus, the disjunction operator is appropriate to evaluate an opinion on a system. In the following, we show how to build an opinion on a system when *(i)* the system has only independent paths and *(ii)* the system has dependent paths.

Independent Paths. Let $\{\sigma_1, \sigma_2, \ldots, \sigma_m\}$ be the paths that enable a user P to achieve an activity. The user's opinion on the paths $\{\sigma_i\}_{i \in \{1..m\}}$ for an activity are denoted by $O_{\sigma_i} = (b_{\sigma_i}, d_{\sigma_i}, u_{\sigma_i}, a_{\sigma_i})$. The user opinion on the system α for achieving the activity, denoted by $O_\alpha = (b_\alpha, d_\alpha, u_\alpha, a_\alpha)$ can be derived by the disjunction of P's opinions in $\{\sigma_i\}_{i \in \{1..m\}}$. Thus, $O_\alpha = \bigvee \{O_{\sigma_i}\}_{i \in \{1..m\}}$. Given Relation (10), we obtain the following generalization: $O_{\alpha = \{\sigma_1, \ldots, \sigma_m\}} =$

$$
\begin{cases}
b_{\alpha = \{\sigma_1, \ldots, \sigma_m\}} = b_{\bigvee \{\sigma_i\}} = 1 - \prod_{i=1}^{m} (1 - b_{\sigma_i}) \\
d_{\alpha = \{\sigma_1, \ldots, \sigma_m\}} = d_{\bigvee \{\sigma_i\}} = \prod_{i=1}^{m} d_{\sigma_i} \\
u_{\alpha = \{\sigma_1, \ldots, \sigma_m\}} = u_{\bigvee \{\sigma_i\}} = \prod_{i=1}^{m} (d_{\sigma_i} + u_{\sigma_i}) - \prod_{i=1}^{m} (d_{\sigma_i}) \\
a_{\alpha = \{\sigma_1, \ldots, \sigma_m\}} = a_{\bigvee \{\sigma_i\}} = \frac{\prod_{i=1}^{m}(d_{\sigma_i} + u_{\sigma_i}) - \prod_{i=1}^{m}(d_{\sigma_i} + u_{\sigma_i} - u_{\sigma_i} a_{\sigma_i})}{\prod_{i=1}^{m}(d_{\sigma_i} + u_{\sigma_i}) - \prod_{i=1}^{m}(d_{\sigma_i})}
\end{cases}
\tag{13}
$$

Again, the proofs of Relation (13) are available in [3].

Dependent Paths. As we know, in subjective logic, as in probabilistic logic, the conjunction is not distributive over the disjunction. In SOCIOTRUST, this problem has been resolved by using conditional probability. As there is not a similar formalism in subjective logic, for evaluating trust in a system we propose to transform a graph having dependent paths to a graph having independent paths. Figure 14 illustrates this transformation. The left side of this figure shows a graph that has three dependent paths. The dependent paths are[7]: $\sigma_1 = \{A, B, C\}$, $\sigma_2 = \{A, E, F\}$ and $\sigma_3 = \{D, E, F\}$. The common nodes are A, E and F. For instance, A is a common node between σ_1 and σ_2. In that transformation, A is duplicated in A_1 and A_2, such that in the new graph, $A_1 \in \sigma'_1 = \{A1, B, C\}$, and $A_2 \in \sigma'_2 = \{A2, E, F\}$, so is the case for the nodes E and F. The right part of Fig. 14 shows the new graph after duplicating the common nodes. The new graph contains the paths $\sigma'_1 = \{A1, B, C\}$, $\sigma'_2 = \{A2, E1, F1\}$ and $\sigma'_3 = \{D, E2, F2\}$. Once this transformation is made, we can apply the Relations (12) and (13). To do so, we propose the following methods that are shown in Algorithms 1 and 2. Notice that lines 1–5 of these algorithms transform the graph.

Copy. In this method, once the graph is transformed to obtain independent paths, we associate the opinion on the original node to the duplicated nodes. This method is based on the idea that the new produced path σ' maintains the same opinion of the original path σ. In this case $O_{\sigma_1} = O_{\sigma'_1}$ and $O_{\sigma_2} = O_{\sigma'_2}$.

---

[7] We recall that the person, the data instance, and the data are not considered in paths of the DAG.

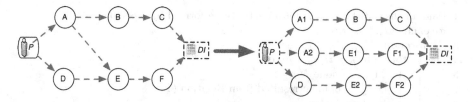

Fig. 14. Graph transformation.

```
 1  Find all the paths σ_{i:i∈[1..n]} for an activity performed by a person
 2  foreach σ_{i:i∈[1..n]} do
 3      foreach N_{j:j∈[1..length(σ_i)]} ∈ σ_i do
 4          foreach k ≠ i: N_j ∈ σ_k do
 5              Create a node N_{ik}
 6              O_{N_{ik}} ← O_{N_j}
 7              Replace N_j by N_{ik} in σ_k
 8          end
 9      end
10  end
```

Algorithm 1. Copy algorithm.

Split: In this method, once the graph is transformed to obtain independent paths, in order to maintain the opinion on the global system, we split the opinion on the original dependent node into independent opinions, such that their disjunction produces the original opinion. Formally speaking, if node A is in common between σ_1 and σ_2, and the opinion on A is O_A, A is duplicated into $A_1 \in \sigma_1'$ and $A_2 \in \sigma_2'$ and the opinion O_A is split into O_{A_1} and O_{A_2}, where O_{A_1} and O_{A_2} satisfy the following relations: $O_{A_1} = O_{A_2}$ and $O_{A_1} \vee O_{A_2} = O_A$. The following is the relation of splitting an opinion into n independent opinions.

$$\bigwedge \begin{cases} O_{A_1} \vee O_{A_2} \vee \ldots \vee O_{A_n} = O_A \\ O_{A_1} = O_{A_2} = \ldots = O_{A_n} \end{cases} \Rightarrow$$

$$\begin{cases} b_{A1} = b_{A2} = \ldots = b_{A_n} = 1 - (1 - b_A)^{\frac{1}{n}} \\ d_{A1} = d_{A2} = \ldots = d_{A_n} = d_A^{\frac{1}{n}} \\ u_{A1} = u_{A2} = \ldots = u_{A_n} = (d_A + u_A)^{\frac{1}{n}} - d_A^{\frac{1}{n}} \\ a_{A1} = a_{A2} = \ldots = a_{A_n} = \frac{(1-b_A)^{\frac{1}{n}} - (1 - b_A - a_A u_A)^{\frac{1}{n}}}{(d_A + u_A)^{\frac{1}{n}} - d_A^{\frac{1}{n}}} \end{cases} \tag{14}$$

Proofs of Relations (14) are provided in [3].

4.3   Experimental Evaluation

In this section, we compare **Copy** and **Split** to a modified version of an approach of the literature named TNA-SL [21]. The latter approach is based on

1 Find all the paths $\sigma_{i:i\in[1..n]}$ for an activity performed by a person
2 **foreach** $\sigma_{i:i\in[1..n]}$ **do**
3 | **foreach** $N_{j:j\in[1..length(\sigma_i)]} \in \sigma_i$ **do**
4 | | **foreach** $k \neq i$: $N_j \in \sigma_k$ **do**
5 | | | Create a node N_{ik}
6 | | | $O_{N_{ik}} \leftarrow$ opinion resulted from Relation (14)
7 | | | Replace N_j by N_{ik} in σ_k
8 | | **end**
9 | **end**
10 **end**

Algorithm 2. Split algorithm.

simplifying the graph by deleting the dependent paths that have high value of uncertainty, then, trust is propagated. In our work, trust is not propagated and a comparison to a propagation approach has no sense. Thus, we modify TNA-SL such that trust evaluation is made by applying Relations (12) and (13) introduced in Sect. 4.2. We call this method "modified TNA-SL", denoted **mTNA** in the following.

We present different experiments, their results, analysis and interpretation. The main objectives are *(i)* to compare the proposed methods and evaluating their accuracy and *(ii)* to confront this approach with real users. The first two experiments are related to the first objective while the third experiment is devoted to the second objective. Next sections present the different experiments, their results, and analysis.

Comparing the Proposed Methods. To tackle the first objective, we experiment with a graph that contains only independent paths. The three methods, **mTNA**, **Copy** and **Split** give the same exact results as expected because the three of them follow the same computational model when graphs contain only independent paths. Then, we experiment on a graph that has relatively high rate of common nodes and dependent paths. 75 % of the paths of the chosen graph are dependent paths and 60 % of nodes are common nodes.

In our experiments, random opinions $O_N = (b_N, d_N, u_N, a_N)$ are associated to each node, and the opinion's probability expectation value of the graph, $\mathbb{E}(O_\alpha) = b_\alpha + a_\alpha u_\alpha$ is computed using the three methods, **mTNA**, **Copy** and **Split**. This experiment is repeated 50 times where each time represents random opinions of a person associated to the different nodes that compose the graph. We analyze the opinion's probability expectation values of the graph, $\mathbb{E}(O_\alpha) = b_\alpha + a_\alpha u_\alpha$ and not all the opinion parameters $O_\alpha = (b_\alpha, d_\alpha, u_\alpha, a_\alpha)$ for simplicity.

Figure 15 shows obtained results. We notice that the three methods almost have the same behavior, when the $\mathbb{E}(O_\alpha)$ increases in one method, it increases in the other methods, and vice versa. We also observe some differences among the three methods that are not always negligible like in experience 9 and 40 in Fig. 15. This observation leads us to the question: *which of these methods*

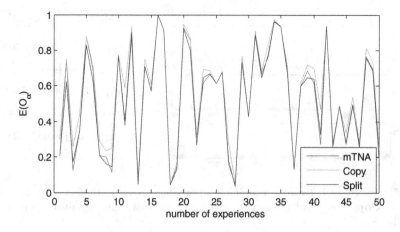

Fig. 15. Value of the probability expectation for 50 persons using the three methods **mTNA**, **Copy** and **Split**.

give the most accurate results? To evaluate the accuracy of **Split**, **Copy** and **mTNA**, we conduct the next experiments.

Studying the Accuracy of the Proposed Methods. SOCIOTRUST that uses theory of probability to evaluate trust in a system, has the advantages that it has no approximations in case there are dependent paths thanks to conditional probability (*cf.* Sect. 3). Thus it works perfectly if users are sure of their judgments of trust, *i.e.*, the values of uncertainty are equal to 0.

Subjective logic is equivalent to traditional probabilistic logic when $b + d = 1$ such that the value of uncertainty is equal to 0. When $u = 0$, the operations in subjective logic are directly compatible with the operations of the traditional probability. In this case the value of $\mathbb{E}(O) = b + au = b$ corresponds to the probability value.

Since SOCIOTRUST is based on probability theory, the obtained results by applying subjective logic if $u = 0$ should be equal to the ones using probability theory. We can evaluate the accuracy of the proposed methods by setting $u = 0$ and comparing the value of $b_\alpha = \mathbb{E}(O_\alpha)$ resulted from applying the three methods to the trust value obtained by applying SOCIOTRUST.

The experiments are conducted on the graph used in Fig. 15. Random opinions $O_N = (b_N, d_N, 0, a_N)$ are associated to each node, and the probability expectation of the graph $\mathbb{E}(O_\alpha) = b_\alpha + a_\alpha u_\alpha = b_\alpha$ is computed. The notations $T_{ST}, T_{\text{mTNA}}, T_{\text{Copy}}, T_{\text{Split}}$ respectively denote system's trust value resulting from applying SOCIOTRUST and system's opinion probability expectation resulting from applying **mTNA**, **Copy**, and **Split**.

To compare T_{ST} to $T_{\text{mTNA}}, T_{\text{Copy}}$, and T_{Split}, we simply compute the subtractions between them *i.e.*, $T_{ST} - T_{\text{mTNA}}, T_{ST} - T_{\text{Copy}}, T_{ST} - T_{\text{Split}}$. The average of each of the previous values are computed through 10,000 times to obtain a

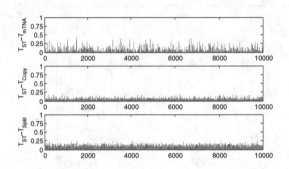

Method	Avg	SD
$\mid T_{ST} - T_{\mathrm{mTNA}} \mid$	0.024	0.045
$\mid T_{ST} - T_{\mathrm{Copy}} \mid$	0.014	0.020
$\mid T_{ST} - T_{\mathrm{Split}} \mid$	0.032	0.037

Fig. 16. Difference between the opinion's probability expectation of a graph using **mTNA**, **Copy**, and **Split** when $u = 0$ and the trust value resulting from using SOCIOTRUST.

reliable value. The standard deviation (SD) is also computed to show how much variation from the average exists in the three cases. Figure 16 shows obtained results.

As we notice from Fig. 16, **Copy** is the method that gives the closest results to SOCIOTRUST, the average of the difference of its result when $u = 0$ and the result of traditional probability over 10,000 times is equal to 0.014, which is an indication that this method gives the nearest result to the exact result and its average error rate is around 1.4 %. **Copy** shows the most convincing result, with a standard deviation equals to 0.02.

The average error rate of **mTNA** (2.4 %) is less than **Split** (3.2 %), but the standard deviation of **mTNA** is 0.045 where in **Split**, it is 0.037. That means that in some cases, **mTNA** can give results that are farther than **Split** from the exact results. Thus, **Split** shows a more stable behavior than **mTNA**.

The objective of this experiment is not criticizing the proposed methods in the literature for the problem of dependent paths. These methods are proposed to deal with the problem of trust propagation through a graph, whereas, in our work we focus on evaluating trust towards the whole graph. The employed operators in our case are different from the employed operators in trust propagation. TNA-SL or any proposed method in the literature can work properly in their context.

In this experiment, we show that **Copy** is the method the more adaptable to be used with respect to the context of our work. Extensive simulations on different types of graphs are provided in [3] and follow the same behavior presented above.

Social Evaluation (A Real Case). In this experiment we use the SVN system of the LINA research laboratory introduced in Sect. 3.4. Since subjective logic is not used yet in real applications, users are not used to build an opinion directly using this logic. We build these opinions ourselves from users' positive or negative observations as it is proposed in [19]. To do that, a survey is executed to collect the observations of LINA users about the nodes. The proposed questions collect information about the user's usage of a node and the quantity of using it and their observations. A local opinion on each entity is built for each user. The opinion and the opinion's probability expectation of the system are then computed using **Copy** for each user. The results are shown in Table 7.

We asked each user for a feedback about their opinion on the nodes and in the system. We were glad to notice that LINA users were satisfied of the obtained results, whereas when using SocioTrust (*cf.* Sect. 3.4), 25 % of users were not satisfied. In the latter approach, when users do not have enough knowledge about a node, they assign the value 0.5, that they consider as neutral value. That leads to incorrect inputs that produce low trust values in a system. In SubjectiveTrust, uncertainties are expressed in the opinions on the nodes and computing an opinion on a system is made considering these uncertainties. That shows that, in uncertain environments, it is more suitable to use subjective logic than probabilistic metrics for trust evaluations.

5   Related Works

The state of the art of this work has two parts, the first one concerns the system modeling (Sect. 5.1) and the second one concerns the evaluation of trust (Sect. 5.2).

5.1   System Modeling

At the beginning of this study, we searched for methodologies that could help us answer questions posed in Sect. 2 and we found interesting approaches in the domain of Enterprise Architecture (EA). In EA, the term "Enterprise" expresses the whole complex, socio-technical system [14], including people, information and technology. A widely known definition of an EA is: *"An Enterprise Architecture is a rigorous description of the structure of an enterprise, which comprises enterprise components, the properties of those components, and the relationships between them"*. The goal of this description is translating the business vision into models. To do that, analytical techniques are used to formalize an enterprise. This allows to produce models that describe the business processes, people organization, information resources, software applications and business capabilities within an enterprise. These models provide the keys that enable the enterprise evolution. Therefore, humans, technical resources, business information, enterprise goals, processes, the roles of each entity in an enterprise, and the organizational structures should be included in this description.

EA is very complicated and large [32]. To manage this complexity, EA Frameworks provide methods and tools that allow to produce enterprise models. Many frameworks have appeared and we studied two of the most used, The Open Group Architecture Framework (TOGAF) [18,23] and the OBASHI Business & IT methodology and framework [34]. More details about these frameworks are available in [3].

TOGAF follows the standard of modeling in four layers: (1) the technology layer, (2) the application layer, (3) the data layer, and (4) the business layer. It defines a metamodel that allows to formalize an enterprise and produce models (diagrams, catalogs and matrices) for the company stakeholders. The metamodel allows to define a formal structure of the components within an architecture like an actor, a role, a data entity, an application, and a business service. Besides the components, the metamodel defines the relationships between these components like an actor belongs to an organization unit or a role is assumed by an actor. TOGAF is a very rich and powerful framework. It produces a set of graphs represented by *diagrams* but none of them can be useful for our needs, *i.e.,* a representation of an activity achieved through a system for a given user. Besides that, obtaining TOGAF diagrams is a complex procedure that needs an expert person. This complexity and the high economical cost of TOGAF leads us to exclude this framework.

The OBASHI framework provides a tool for capturing, illustrating and modeling the relationships of dependency and the dataflows between business and IT environment in a business context. OBASHI does not have a specific metamodel to formalize the enterprise components, instead, it proposes a classification for the components which should be located in the layer that corresponds to their type. OBASHI has six layers: **O**wnership, **B**usiness processes, **A**pplications, **S**ystems, **H**ardware, and **I**nfrastructure. The OBASHI relationships describe the relations between the components, which follow the OBASHI rules. This model allows to create the business and IT diagrams (B&IT) and the dataflow, which are the main output of the OBASHI tool that helps the enterprise to develop its work and understand its needs. Despite its simplicity, OBASHI does not answer our needs. The B&IT Diagram and the data flow present a dependency graph that allows to find the sequences of the dependencies relations between the entities in an enterprise. In our work, the resulting model should represent an activity achieved through a system by a given user, more precisely, the model should contain the entities this user depends on to perform an activity and not the flow of dependencies between entities in a system.

In general, what mainly distinguishes SOCIOPATH from EA, is the social world that focuses on the persons who participate to the system. Instead of the social world, EA presents the business layer, which is mainly introduced by the component organization or organization unit. Hence, the analysis of the information in SOCIOPATH focuses on the needs of the person who uses a system including her social, digital and physical dependences. Whereas, in EA, the analysis of the information focuses on the needs of an enterprise including ameliorating its performance, choosing the best person for a particular task, *etc.*

The Business Process Modeling and Notation (BPMN) [11,33], is a standard to model business process mainly in the early phases of system development. To build diagrams, BPMN provides four categories of graphical elements. (1) Flow Objects, represent all the actions which can happen inside a business process determining its behavior. They consist of Events, Activities and Gateways. (2) Connecting Objects, provide three different ways of connecting

Table 7. Users' opinions in the system for the activity "a user access a file on the SVN" at LINA laboratory.

	O_A (b,d,u,a)	O_B (b,d,u,a)	O_G (b,d,u,a)	O_D (b,d,u,a)	O_E (b,d,u,a)	O_F (b,d,u,a)	O_G (b,d,u,a)	O_α (b,d,u,a)	$\mathbb{E}(O_\alpha)$
P_1	(1,0,0,0.5)	(0,0,1,0.5)	(1,0,0,0.5)	(0.99,0.01,0,0.5)	(0.83,0,0.17,0.5)	(0.99,0.01,0,0.5)	(0.83,0,0.17,0.5)	(0.6820,0,0.3810,0.8389)	0.9488
P_2	(1,0,0,0.5)	(1,0,0,0.5)	(1,0,0,0.5)	(1,0,0,0.5)	(0.83,0,0.17,0.5)	(1,0,0,0.5)	(1,0,0,0.5)	(1,0,0,−)	1
P_3	(0.99,0.01,0,0.5)	(0,0,1,0.5)	(1,0,0,0.5)	(0.99,0.01,0,0.5)	(0,0,1,0.5)	(0.99,0.01,0,0.5)	(0.6,0,0.4,0.5)	(0,0,1,0.7753)	0.7753
P_4	(1,0,0,0.5)	(0,0,1,0.5)	(1,0,0,0.5)	(1,0,0,0.5)	(0.83,0,0.17,0.5)	(0.99,0.01,0,0.5)	(0.96,0,0.04,0.5)	(0.7888,0,0.2112,0.8604)	0.9705
P_5	(0.99,0.01,0,0.5)	(0,0,1,0.5)	(1,0,0,0.5)	(0.99,0.01,0,0.5)	(0,0,1,0.5)	(1,0,0,0.5)	(0.5,0,0.5,0.5)	(0,0,1,0.7500)	0.75
P_6	(0.99,0.01,0,0.5)	(0,0,1,0.5)	(1,0,0,0.5)	(0.9,0.1,0,0.5)	(0.5,0.5,0,0.5)	(1,0,0,0.5)	(0.6,0,0.4,0.5)	(0.2970,0,0.7030,0.7755)	0.8422
P_7	(0.99,0.01,0,0.5)	(0,0,1,0.5)	(1,0,0,0.5)	(0.9,0.1,0,0.5)	(0.83,0,0.17,0.5)	(1,0,0,0.5)	(1,0,0,0.5)	(0.8217,0,0.1783,0.8522)	0.9736
P_8	(1,0,0,0.5)	(0,0,1,0.5)	(1,0,0,0.5)	(0.9,0.1,0,0.5)	(0.5,0.5,0,0.5)	(1,0,0,0.5)	(1,0,0,0.5)	(0.5000,0,0.5000,0.8625)	0.9313
P_9	(1,0,0,0.5)	(1,0,0,0.5)	(1,0,0,0.5)	(0.95,0.05,0,0.5)	(0,0,1,0.5)	(0.99,0.01,0,0.5)	(0.96,0,0.04,0.5)	(0.9956,0,0.0044,0.7583)	0.9989
P_{10}	(0.99,0.01,0,0.5)	(0,0,1,0.5)	(1,0,0,0.5)	(0.9,0.1,0,0.5)	(0,0,1,0.5)	(0.8,0.2,0,0.5)	(0.98,0,0.02,0.5)	(0,0.0047,0.9953,0.7972)	0.7934
P_{11}	(0.99,0.01,0,0.5)	(0,0,1,0.5)	(1,0,0,0.5)	(0.95,0.05,0,0.5)	(0.72,0,0.28,0.5)	(0.99,0.01,0,0.5)	(0.96,0,0.04,0.5)	(0.6774,0.0001,0.3225,0.8489)	0.9512
P_{12}	(1,0,0,0.5)	(0,0,1,0.5)	(1,0,0,0.5)	(0.95,0.05,0,0.5)	(0.83,0,0.17,0.5)	(0.95,0.05,0,0.5)	(1,0,0,0.5)	(0.7885,0,0.0001,0.2114,0.8301)	0.9640
P_{13}	(1,0,0,0.5)	(0,0,1,0.5)	(1,0,0,0.5)	(0.95,0.05,0,0.5)	(0.83,0,0.17,0.5)	(0.95,0.05,0,0.5)	(0.83,0,0.17,0.5)	(0.6545,0.0001,0.3545,0.8110)	0.9346
P_{14}	(1,0,0,0.5)	(0,0,1,0.5)	(1,0,0,0.5)	(0.99,0.01,0,0.5)	(0.72,0,0.28,0.5)	(0.99,0.01,0,0.5)	(0.72,0,0.28,0.5)	(0.5132,0,0.4868,0.8186)	0.9117
P_{15}	(1,0,0,0.5)	(1,0,0,0.5)	(1,0,0,0.5)	(0.99,0.01,0,0.5)	(0.72,0,0.28,0.5)	(0.99,0.01,0,0.5)	(0.83,0,0.17,0.5)	(0.9870,0,0.0130,0.8492)	0.9980
P_{16}	(0.99,0.01,0,0.5)	(0,0,1,0.5)	(1,0,0,0.5)	(0.9,0.1,0,0.5)	(0.5,0.5,0,0.5)	(1,0,0,0.5)	(0.72,0,0.28,0.5)	(0.3564,0,0.6436,0.8011)	0.8719
P_{17}	(1,0,0,0.5)	(0,0,1,0.5)	(1,0,0,0.5)	(0.99,0.01,0,0.5)	(0.83,0,0.17,0.5)	(1,0,0,0.5)	(0.83,0,0.17,0.5)	(0.6889,0,0.3111,0.8447)	0.9517
P_{18}	(1,0,0,0.5)	(0,0,1,0.5)	(1,0,0,0.5)	(0.99,0.01,0,0.5)	(0.83,0,0.17,0.5)	(1,0,0,0.5)	(1,0,0,0.5)	(0.8300,0,0.1700,0.8737)	0.9785
P_{19}	(1,0,0,0.5)	(1,0,0,0.5)	(1,0,0,0.5)	(0.99,0.01,0,0.5)	(0,0,1,0.5)	(1,0,0,0.5)	(0.72,0,0.28,0.5)	(0.9196,0,0.0804,0.8525)	0.9811
P_{20}	(1,0,0,0.5)	(0,0,1,0.5)	(1,0,0,0.5)	(0.95,0.05,0,0.5)	(0,0,1,0.5)	(1,0,0,0.5)	(0.83,0,0.17,0.5)	(0,0,1,0.8836)	0.8336

various objects: Sequence Flow, Message Flow and Association. (3) Swimlanes, provides the capability of grouping modeling elements. Swimlanes have two elements through which modelers can group other elements: Pools and Lanes. And (4), Artifacts are used to provide additional information about Process that does not affect the flow. They are: Data Object, Group and Annotation. BPMN is a complete standard oriented to business users, business analysts, business staff and technical developers. For our specific needs, to define relations (of control, access, provides, *etc.*) among all the entities a (final) user depends on to achieve an activity, building our approach over BPMN is difficult because we have different focus and semantics. Mapping between this standard and our proposal may exist, but we have not investigated this direction.

5.2   Trust Evaluation

There are many approaches for evaluating trust in the literature [2,8,15,20,27] and several interesting surveys analyze them [7,16,22,37]. The approaches closest to our work are those oriented to graphs [1,17,20,21,26,31]. They are especially used in social networks where the main idea of trust derivation is to propagate it between two nodes in a graph that represents the social network. A social network is a social structure composed of a set of persons (individuals or organizations) and a set of relations among these persons. It can be represented as a graph where the nodes are the persons and the edges are the relations between them. Trust between two persons in a social network can be evaluated based on this graph where the source node is the trustor, the target node is the trustee and the other nodes are the intermediate nodes between the trustor and the trustee. Values are associated to the edges to represent the trust value attributed by the edge source node towards the edge target node. To evaluate trust in a target node in a graph, in general, the following two steps are considered: *(1)* trust propagation through a path and *(2)* trust propagation through a graph employing different metrics and operators. Figure 8 (page 17) shows an example of trust relationships in a social network. For instance, B trusts C with the value 0.8.

Trust propagation focuses on finding a trust value from a person towards another given person through the multiple paths that relate them. For instance, in Fig. 8, how much A trusts E, knowing that there are two paths that relate A with E, and that paths have nodes in common?

In graph-based trust approaches, this problem is either ignored [31], either simple solutions are proposed like choosing one path in a graph [26], or removing the paths that are considered unreliable [17,21]. In [21], Jøsang *et al.* propose a method based on graph simplification and trust derivation with subjective logic named, Trust Network Analysis with Subjective Logic (TNA-SL). They simplify a complex trust graph into a graph having independent paths by removing the dependent paths that have a high value of *uncertainty*. The problem of the previous solution is that removing paths from a graph could cause loss of information. To solve this problem, in another work [20] Jøsang *et al.*, propose to transform a graph that has dependent paths to a graph that has independent paths by duplicating the edges in common and splitting the associated opinions to them.

In this work we propose two approaches that deal with the problem of dependent paths. What differentiates our approaches from those of the literature is that we search to evaluate trust in a system as a whole for an activity and from the point of view of a person. In addition, we argue that the trust in a system depends on its architecture, more precisely, on the way the implicit and explicit entities, which the users depends on to do their activities, are organized.

Comparing trust approaches is hard, an approach is better if its produced trust values are lower (or higher) than another? Which is the reference to say what is a good trust value? That is why, in our experiments we make an effort to confront our proposed approaches to real users (*cf.* Sect. 3.4, page 23 and Sect. 4.3, page 33).

6   Conclusion and Perspectives

Digital activities are achieved everyday by users through different systems. When users need to choose a system for a particular activity, they evaluate it considering many criteria like QoS, economical aspects, *etc.* This paper enlightens some aspects of digital systems to improve users' expectations. The aspects we focused on are the user's digital and social dependences in a system for an activity, their degrees and the level of a user's trust towards the used system. To realize this approach, we fixed two main objectives:

1. Proposing a model that formalizes a system considering the different entities that compose it (physical, digital or social entities) and the relations between them.
2. Evaluating trust in a system for an activity based on this model.

In this paper, we proposed SOCIOPATH, a simple model that allows to formalize the entities in a system and the relations between them. In this contribution, we observed that the entities that compose a digital system can be digital, physical or human entities. We defined a model that formalizes all these entities and the relations between them. We provided this model with the rules that discover some implicit relations in a system and enriched it with definitions that illustrate some main concepts about the used system. SOCIOPATH allows to answer the user of some main questions that concern her used system.

Trust works in the literature focus on one granularity of trust; trusting a person, a product, a resource, *etc.* That reflects one entity in a used system. Trusting a system as a composition of a set of entities and relations between them has not been studied deeply.

By focusing on trust works existing in the literature, one direction drew our attention. This direction is trust propagation in social networks. This approach aims to propagate trust between two nodes in a graph that represents a social network. The propagated trust value results from combining trust values through this graph.

From SOCIOPATH models, we can obtain a directed acyclic graph (DAG) where nodes represent a set of entities that plays a role for achieving the users' activity and the set of edges represents the paths a user follows to achieve her activity.

Based on this DAG we proposed two approaches to evaluate trust in a system for an activity. The first one, SOCIOTRUST, is based on probability theory. It can be used in the field of full-knowledge environments. In presence of uncertainty, the second approach based on subjective logic, SUBJECTIVETRUST, is more suitable. The necessary relations and algorithms for combining the trust values towards the entities in the DAG have been provided and proved, and experiments have been conducted to validate these approaches.

All the evaluations of trust in a system we propose in this article are static. This is a limitation. To achieve a better comprehension of trust in a system and the parameters that can influence it, it will certainly be necessary to consider the evolution of trust over the time. We are convinced that such understanding is a a challenging issue. For this purpose, it is also necessary to compare synthetic trust and real trust of a user. Yet, to the best of our knowledge, there is no method to measure a distance or similarity between an assessment of confidence and the one felt by users. It is certainly possible to build on work already carried out in the fields of Information Retrieval or Social Sciences, but this is a problem we encountered without providing a complete answer. Indeed in our work, we collected users' impressions through a form and showed they feel closer to a proposal than the other. However, a general method of comparison and measurement between an assessment of the trust and the trust really felt remains to build.

It is also interesting to note that SOCIOPATH is not restricted to trust evaluation. Indeed, pointing out accesses and controls relations within an architecture is also related to privacy. Thus, as future work, it could be interesting to use SOCIOPATH to study the compliance of system with users privacy policies.

References

1. Agudo, I., Fernandez-Gago, C., Lopez, J.: A model for trust metrics analysis. In: Furnell, S.M., Katsikas, S.K., Lioy, A. (eds.) TrustBus 2008. LNCS, vol. 5185, pp. 28–37. Springer, Heidelberg (2008)
2. Al-Bakri, M., Atencia, M., Rousset, M.-C.: TrustMe, i got what you mean!. In: ten Teije, A., Völker, J., Handschuh, S., Stuckenschmidt, H., d'Acquin, M., Nikolov, A., Aussenac-Gilles, N., Hernandez, N. (eds.) EKAW 2012. LNCS, vol. 7603, pp. 442–445. Springer, Heidelberg (2012)
3. Alhadad, N.: Bridging the Gap between Social and Digital Worlds: System Modeling and Trust Evaluation. Ph.D. thesis, Université de Nantes, France (2014)
4. Alhadad, N., Busnel, Y., Serrano-Alvarado, P., Lamarre, P.: Trust evaluation of a system for an activity with subjective logic. In: Eckert, C., Katsikas, S.K., Pernul, G. (eds.) TrustBus 2014. LNCS, vol. 8647, pp. 48–59. Springer, Heidelberg (2014)
5. Alhadad, N., Lamarre, P., Busnel, Y., Serrano-Alvarado, P., Biazzini, M., Sibertin-Blanc, C.: SocioPath: Bridging the gap between digital and social worlds. In: Proceedings of the 23rd International Conference on Database and Expert Systems Applications (DEXA), pp. 497–505 (2012)

6. Alhadad, N., Serrano-Alvarado, P., Busnel, Y., Lamarre, P.: Trust evaluation of a system for an activity. In: Furnell, S., Lambrinoudakis, C., Lopez, J. (eds.) Trust-Bus 2013. LNCS, vol. 8058, pp. 24–36. Springer, Heidelberg (2013)
7. Bloehdorn, S., Sure, Y.: Kernel methods for mining instance data in ontologies. In: Aberer, K., Choi, K.-S., Noy, N., Allemang, D., Lee, K.-I., Nixon, L.J.B., Golbeck, J., Mika, P., Maynard, D., Mizoguchi, R., Schreiber, G., Cudré-Mauroux, P. (eds.) ASWC 2007 and ISWC 2007. LNCS, vol. 4825, pp. 58–71. Springer, Heidelberg (2007)
8. Atencia, M., Al-Bakri, M., Rousset, M.-C.: Trust in networks of ontologies and alignments. Knowl. Inf. Syst. 42(2), 1–27 (2015)
9. Aurrecoechea, C., Campbell, A.T., Hauw, L.: A survey of QoS architectures. Multimedia Syst. 6(3), 138–151 (1996)
10. Blau, P.: Exchange and Power in Social Life. John Wiley and Sons, New York (1964)
11. Chinosi, M., Trombetta, A.: BPMN: an introduction to the standard. Comput. Stan. Interfaces 34(1), 124–134 (2012)
12. Emerson, R.M.: Power-dependence relations. Am. Sociol. Rev. 27(1), 31–41 (1962)
13. Gambetta, D.: Can we trust trust. In: Trust: Making and Breaking Cooperative Relations, vol. 13, pp. 213–237. Department of Sociology, University of Oxford (2000)
14. Giachetti, R.E.: Design of Enterprise Systems: Theory, Architecture, and Methods. CRC Press, Boca Raton Florida (2010)
15. Golbeck, J.: Computing and Applying Trust in Web-based Social Networks. Ph.D. thesis, Department of Computer Science, University of Maryland (2005)
16. Golbeck, J.: Trust on the world wide web: a survey. Found. Trends web Sci. 1(2), 131–197 (2006)
17. Golbeck, J., Hendler, J.A.: Inferring binary trust relationships in web-based social networks. ACM Trans. Internet Technol. 6(4), 497–529 (2006)
18. Harrison, R.: TOGAF Version 8.1. Van Haren Publishing, New York (2007)
19. Jøsang, A.: A logic for uncertain probabilities. Uncertainty, Fuzziness Knowl.-Based Syst. 9(3), 279–311 (2001)
20. Jøsang, A., Bhuiyan, T.: Optimal trust network analysis with subjective logic. In: Proceeding of the 2nd International Conference on Emerging Security Information, Systems and Technologies (SECURWARE), pp. 179–184 (2008)
21. Jøsang, A., Hayward, R., Pope, S.: Trust network analysis with subjective Logic. In: Proceedings of the 29th Australasian Computer Science Conference (ACSC), pp. 85–94 (2006)
22. Jøsang, A., Ismail, R., Boyd, C.: A survey of trust and reputation systems for online service provision. Decis. Support Syst. 43(2), 618–644 (2007)
23. Josey, A.: TOGAF Version 9: A Pocket Guide, 2nd edn. Van Haren Publishing (2009)
24. Li, L., Wang, Y.: Subjective trust inference in composite services. In: Proceedings of the 24th Conference on Artificial Intelligence (AAAI) (2010)
25. Li, L., Wang, Y.: A subjective probability based deductive approach to global trust evaluation in composite services. In: Proceedings of the 9th IEEE International Conference on Web Services (ICWS), pp. 604–611 (2011)
26. Liu, G., Wang, Y., Orgun, M., Lim, E.: Finding the optimal social trust path for the selection of trustworthy service providers in complex social networks. IEEE Trans. Serv. Comput. 6(2), 152–167 (2011)
27. Marsh, S.P.: Formalising Trust as a Computational Concept. Ph.D. thesis, Department of Mathematics and Computer Science, University of Stirling (1994)

28. Mcknight, D.H., Chervany, N.L.: The Meanings of Trust. Technical report, University of Minnesota, Carlson School of Management (1996)
29. Molm, L.: Structure, Action, and Outcomes: The Dynamics of Power in Social Exchange. American Sociological Association Edition (1990)
30. Moyano, F., Fernandez-Gago, C., Lopez, J.: A conceptual framework for trust models. In: Fischer-Hübner, S., Katsikas, S., Quirchmayr, G. (eds.) TrustBus 2012. LNCS, vol. 7449, pp. 93–104. Springer, Heidelberg (2012)
31. Richardson, M., Agrawal, R., Domingos, P.: Trust management for the semantic web. In: Fensel, D., Sycara, K., Mylopoulos, J. (eds.) ISWC 2003. LNCS, vol. 2870, pp. 351–368. Springer, Heidelberg (2003)
32. Rohloff, M.: Enterprise Architecture-framework and methodology for the design of architectures in the large. In: Proceedings of the 13th European Conference on Information Systems (ECIS), pp. 1659–1672 (2005)
33. The official BMPN Website. http://www.bpmn.org/. Accessed May 2015
34. The official OBASHI Website. http://www.obashi.co.uk/. Accessed May 2015
35. Viljanen, L.: Towards an ontology of trust. In: Katsikas, S.K., López, J., Pernul, G. (eds.) TrustBus 2005. LNCS, vol. 3592, pp. 175–184. Springer, Heidelberg (2005)
36. Yan, Z., Holtmanns, S.: Computer Security, Privacy and Politics: Current Issues, Challenges and Solutions, chapter Trust Modeling and Management: from Social Trust to Digital Trust. IGI Global (2007)
37. Zhang, P., Durresi, A., Barolli, L.: Survey of Trust Management on Various Networks. In: Proceedings of the 5th International Conference on Complex, Intelligent and Software Intensive Systems (CISIS), pp. 219–226 (2011)

Efficient Querying of XML Data Through Arbitrary Security Views

Houari Mahfoud[1]([⊠]) and Abdessamad Imine[2]

[1] Abou-Bekr Belkaïd University, Tlemcen, Algeria
houari.mahfoud@gmail.com
[2] University of Lorraine and INRIA-LORIA, Nancy, France

Abstract. We study the problem of querying virtual security views of XML data that has received a great attention during the past years. A major concern here is that user XPath queries posed on recursive views cannot be rewritten to be evaluated on the underlying XML data. Existing rewriting solutions are based on the non-standard language, "Regular XPath", which makes rewriting possible under recursion. However, query rewriting under Regular XPath can be of exponential size. We show that query rewriting is always possible for arbitrary security views (recursive or not) by using only the expressive power of the standard XPath. We propose a more expressive language to specify XML access control policies as well as an efficient algorithm to enforce such policies. Finally, we present our system, called SVMAX, that implements our solutions and we show that it scales well through an extensive experimental study based on real-life DTD.

Keywords: XML access control · Security views · Materialization · Query rewriting · XPath · Regular XPath · XML databases · Confidentiality and integrity

1 Introduction

In parallel with the rapid growth of the World Wide Web, an increasing amount of data have become available electronically to humans and programs. Such data may be combined from heterogeneous systems based on different data formats, and need to be maintained in a self-describing format to accommodate a variety of ever-evolving business needs. This has led a need for a neutral and flexible way for exchanging data among different devices, systems, and applications. The solution to this problem came with the advent of XML [1,2].

The *eXtensible Markup Language* (XML) is a W3C recommendation that encodes data in a format which can be processed easily and exchanged across multiple platforms. XML has been universally received as the de facto standard for representing and exchanging data. An XML document represents not only base information, but also information about the relationship of data items to each other in the form of the hierarchy (*hierarchical structure*). Moreover, it

© Springer-Verlag Berlin Heidelberg 2015
A. Hameurlain et al. (Eds.): TLDKS XXII, LNCS 9430, pp. 75–114, 2015.
DOI: 10.1007/978-3-662-48567-5_3

can be searched or updated without requiring a static definition of the schema (*schema-less property*). XML brings a number of powerful capabilities to information modeling: (a) *Heterogeneity* (each record can contain different data fields); (b) *Extensibility* (new types of data can be added at will and do not need to be determined in advance); and (c) *Flexibility* (data fields can vary in size and configuration from instance to instance). These features and capabilities have made of XML the most used format for several needs and within various situations:

- *XML-based technologies*: XML has emerged as a critical enabler to various technology initiatives. Service-oriented architectures (*SOA*), enterprise application integration (*EAI*), web services, and standardization efforts in many industries all rely on or make use of XML as an underlying technology.

- *XML-based languages*: XML contributes on the creation of many markup languages for various domains such as *MathML* for mathematic, *CML* for chemistry, *SBML* and *BIOPAX* for biology, *SCORM* for e-learning.

- *Desktop applications*: *Open-office* files, *Ant's Build* files, and *Mac plist* configuration files are all written in XML format.

Specifically, we focus on situations where XML does not serve just as a technology or a configuration model, as explained above, but as a primordial format for data representation and exchange. Such situations are often encountered when the managed data has a *volatile schema* and is *inherently hierarchical* in nature. The properties of XML make it an *unavoidable* and more suitable format for this kind of data. We take first the case of medical data which is often presented with XML. A simple scenario of that is the "*Electronic Health Record (ERH)*[1]", an ongoing national project started in France at 2004, which has as goal to allow each one to access electronically to his own medical data (e.g. personal information, appointments, analysis results, medical and surgical history). Data of two different patients may not have the same rigid structure, e.g. one patient may have some surgical information whereas the other does not have any surgical item. Each department within the hospital may maintain his own volatile and local schema of data, and all schema may be combined to form the global data schema of the hospital. Moreover, the hospital data may be exchanged with other hospitals or laboratories that are not supposed to use the same schema. According to this context, it is much more natural to use XML for local data representation and to ensure efficiently the mapping between the different schema [3,4]. Many XML-based solutions are proposed for managing medical data: the *hData* [5] and *MEDOX* [6] frameworks, the *HL7* standard[2], solutions for interoperability of health-care applications [7], security [8,9] and integration [10,11] of health-care data. The other case occurs with the e-business where XML is an unavoidable standard not only for data representation, but essentially for ensuring interoperability of different systems. For this purpose, many XML-based

---

[1] The original name is the *DMP*, that refers in French to "*Dossier Médical Personnel*".
[2] Available at: http://www.hl7standards.com/.

solutions have been proposed: the *IBM jStart team* [12], the *DITA OASIS Standard* [13], the *ebXML consortium* [14], and the *Oracle* solutions [15] rely all on XML to address needs of managing and publishing business information.

Day to day operations that use XML data need to be easy to use, quick to carry out, and more importantly *safe* from *unauthorized* accesses. For instance, electronic commerce transactions require enforcement of some security constraints ensuring that crucial information will be accessible only to authorized entities. In addition, many organizations (mostly medical and commercial) manipulate *sensitive* information that should be *selectively exposed* to different classes of users based on their *access privileges*. A good example of such sensitive data is the *"EHR"* explained above. All patients' data are stored in a centralized database, and can be accessed totally/partially by different health personnels: nurses, doctors, pharmacists, insurance company staff, etc. Due to the sensitive nature of this data, a security policy is applied that controls access to different parts of the health-care data. For instance, grant to an insurance company a read access that concerns only medication information. The general scenario that can be found in practice is the following. For some XML data there may be multiple user groups which want to query the same data. For these user groups, different access privileges may be imposed, specifying what parts of the data are accessible to the users. The problem of secure XML access is to enforce these privileges. The well-established security specification and enforcement approaches of relational databases cannot be easily adapted for XML databases. This can be explained by the fact that XML has its own properties: an *hierarchical structure*, *schemaless* and *node relationship* properties. Consequently, the problem of secure access to XML data has its own particular flavor and requires specific solutions.

1.1   Motivation

It is increasingly common nowadays to find virtual views used to protect access to data as supported by many database systems (e.g. *Oracle 11g, IBM DB2*). Different models have been proposed that study such kind of protection [16–21]. Most of them deal only with read access rights. Given an XML document T that conforms to a schema D, a security view S is defined that heads some inaccessible information from D. According to S, a schema view D_v is derived first and provided to the user that describes the accessible data (s)he is able to see. Moreover, a virtual data view T_v is extracted that displays only accessible parts of T. XPath [22,23] is the most used language to query such virtual data view. For each XPath query Q posed on T_v, the *query rewriting* principle consists on rewriting Q into another one Q' such that: evaluating Q over T_v yields the same result as the evaluation of Q' over the original document T. Many rewriting algorithms have been proposed during the last decade [16,18,19,21,24,25].

Although a tremendous effort has been done on improving query rewriting over virtual XML views, most of existing algorithms are limited in the sens that they deal only with non recursive schema[3]. We investigate the use of DTD grammar as data schema. Recursive DTDs often arise in practice when specifying for instance (bio)medical and biological data. Examples of such DTDs are GedML and BIOML. The study done in [26] shown that most of the real-world DTDs are recursive. The rewriting process over virtual views becomes more challenging when manipulating recursive DTDs. Specifically, for two *accessible* nodes A and B, there may be some *inaccessible* nodes that connect A with B at the original data, these nodes are hidden in the view and thus B appears as immediate child of A in the virtual data view. Each query A/B must be rewritten to return only accessible B nodes that are either immediate children of some accessible nodes A or connected to them with only inaccessible nodes. Roughly, to rewrite a query A/B it remains to find all the inaccessible paths[4] that connect accessible nodes A with accessible B at the original data. Because of recursion, these paths may lead to an infinite set which cannot be explicitly expressed with the standard XPath. Thus, the query rewriting over recursive views is still an open problem.

For this reason, Fan et al. [17,27] proposed, as extension of their previous work [25], the first algorithm for coping with recursive security views. Their algorithm has been refined later by Groz et al. [18] by considering different types of DTDs and larger class of queries. The key idea behind these three works was to use the Regular XPath language [28] that is more expressive than the standard XPath and offers possibility to define recursive paths by means of the *Kleene star* operator "*". Although Regular XPath ensures query rewriting over arbitrary security views (recursive or non), this process may be costly since rewritten queries may be of exponential size. Regular XPath based investigations cannot be easily applied in practice: no tool exists to evaluate Regular XPath queries. Furthermore, more commercial database systems (e.g. *Oracle 11g, IBM DB2, eXist-db, Sedna*) provide support for the standard XPath to manipulate XML data. Therefore, there is a need for an XPath-based practical solution to secure XML data over arbitrary views.

Given the above, our first motivation at the outset was to develop some *practical* security solutions that can be easily and efficiently integrated within existing systems that provide support for managing XML data. We have focused principally on shortcomings of security-view-based approaches [17,18,24,25], and investigated some practical and efficient solutions to overcome them. This paper is thus a continuation to the important effort done during the two decades to design and implement XML access control models.

1.2   Contributions

An Efficient Approach for Coping with Arbitrary XML Security Views. While, in case of recursive security views, the query rewriting is not always possible over the downward fragment of XPath[5] [17] (class of queries

---

[3] A *recursive* schema has at least an element defined (in)directly in terms of itself.

[4] Paths composed by only inaccessible nodes.

[5] This fragment is more used both in practice and in theory, and several theoretical results have been found around this fragment [29,30].

with *child*-axis, *descendant*-axis, and complex predicates), we show that the expressive power of the standard XPath is sufficient to overcome this rewriting limitation. We extend the access specification language of Fan et al. [25] with new annotation types in order to define compact and more expressive XML access control policies. Then, we show that by extending the downward fragment of XPath with some axes and operators, the query rewriting becomes possible under arbitrary security views (recursive or non). As explained in Sect. 5, our rewriting approach can deal with a larger class of XPath queries that includes downward-axes (*child*, *descendant*), upward-axes (*parent*, *ancestor*). Moreover, it can be easily extended to rewrite horizontal-axes (*preceding*, *following*). We propose finally an efficient algorithm to rewrite XPath queries over arbitrary security views. Compared with the one presented in [17,27], our algorithm uses only the access specification (i.e. the read-access annotations) to rewrite any user query rather than using an auxiliary structure, like automatons, which can be costly or even impracticable in some cases. Moreover, our algorithm runs in linear time in the size of the query.

SVMAX system has been implemented to show the practicality and efficiency of our results. To our knowledge, SVMAX (*Secure and Valid MAnipulation of XML*) is the first system that provides secure handling of XML data over arbitrary views (recursive or non).

Further Contributions. We emphasize that SVMAX implements some other solutions, that are not explained here, but which are based on the results of this paper and then deserve a little discussion to complete the description of the system. We studied the XML access control by considering the operations of the XQuery Update Facility [31]. Our results in this context are based principally on the contribution of this paper. More precisely, we proposed in [32,33] a fine-grained language to specify XML update policies and which overcomes expressiveness limits of existing models [21,34]. Our update specification language is an extension of the read-access specification language that we describe in Sect. 4. SVMAX implements a linear time algorithm to enforce our XML update policies.

As we shall explain, SVMAX provides visual editor that helps the administrator to specify either read and update policies. These policies are enforced through the rewriting modules of the system: *XPath Rewriter* and *XQuery Update Rewriter* to rewrite safely, and w.r.t the corresponding policy, read-access queries and update queries respectively.

The wide use of W3C standards in practice makes of SVMAX a useful system that can be easily integrated, as an API, within commercial database systems. See [35] for more details of the system.

1.3   Outline of the Paper

The remainder of the paper is organized as follows. Section 2 provides essentially background about XML and XPath query language. We explain in Sect. 3 the main problem we tackle throughout this paper. Section 4 presents formal description of our access control model, and especially our access specification language. Policies based on such language are enforced through the rewriting approach explained in Sect. 5. Section 6 presents a brief overview of our system, followed by an extensive experimental study based on real-world DTDs. Related work is reviewed in Sect. 7. Finally, we conclude the paper in Sect. 8.

Additional parts of our contributions (algorithms, proofs,...) can be found on-line at https://tel.archives-ouvertes.fr/tel-01093661/.

2   Preliminaries

We present basic notions and definitions that are used throughout the paper.

2.1   Document Type Definitions

Definition 1 (DTD [1]). *A Document Type Definition (DTD) D is a triple* $(\Sigma, P, Root)$, *where* Σ *is a finite set of* element types; *Root is a distinguished type in* Σ *called the* root type; *and P is a function defining element types such that for any A in* Σ, *P(A) is a regular expression* α, *called the* content model *of A, and defined as follows:*

$$\alpha := str \mid \epsilon \mid B \mid \alpha','\alpha \mid \alpha'|'\alpha \mid \alpha* \mid \alpha+ \mid \alpha?$$

where str *denotes the text type PCDATA,* ϵ *is the empty word,* B *is an element type in* Σ, $\alpha','\alpha$ *denotes concatenation, and* $\alpha'|'\alpha$ *denotes disjunction.* $A \to P(A)$ *refers to the* production rule *of A. For each element type* B *occurring in P(A), we refer to* B *as a* child type *of A and to A as a* parent type *of B. Moreover, P(A) can be defined using the operators '*' (set with zero or more elements), '+' (set with one or more elements), and '?' (optional set of elements). A DTD D is* recursive *if some element type A is defined (in)directly in terms of itself.*

Example 1. We consider the *department* DTD $(\Sigma, P, dept)$ with $\Sigma = \{dept, course, project, cname, takenBy, givenBy, students, scholarship, student, sname, mark, professor, pname, grade, type, private, public, descp, results, result, members, member, name, qualif, theoretical, experimental, sub-project\}$. The production rules of this DTD are defined as follows:

$$
\begin{aligned}
dept &\to (course+, project*) \\
course &\to (cname, takenBy, givenBy) \\
takenBy &\to (students) \\
students &\to (scholarship?, student+) \\
scholarship &\to (student+) \\
student &\to (sname, mark) \\
givenBy &\to (professor+) \\
professor &\to (pname, grade) \\
project &\to (type, descp, results, members, sub-project) \\
type &\to (private \mid public) \\
results &\to (str \mid result)* \\
members &\to (member+) \\
member &\to (name, qualif, (theoretical \mid experimental)*) \\
sub\text{-}project &\to (project*)
\end{aligned}
$$

The element types *private* and *public* are empty, while the remaining element types (e.g. *mark, result*) are text elements. A *dept* element has a list of *course* elements as well as zero or more *project* elements. A *course* consists of *cname* (course name), and lists of *students* and *professor* elements defined via the relations *takenBy* and *givenBy* respectively. A *student* who has registered for the *course* has a name (*sname*), a *mark* and may be part of a *scholarship* program. A *professor* is defined by his name (*pname*) and *grade*. A *project* is presented by its *type* (that can be either *private* or *public*), a description (*descp*), some *results*, and may be composed by zero or more than one projects (through the *sub-project* relation). A *member* of a given *project* is presented by his *name*, a qualification (denoted *qualif* that can be *professor, student, external researcher* etc.), and a list of his contributions (that can be either *theoretical* or *experimental*). Notice that *results* element type has mixed content (combination of text values that serve as comments, and *result* elements). Moreover, *member* element type has complex content, i.e. a sequence container that has the choice container (*theoretical* | *experimental*)*.   □

2.2 XML Documents

We model an XML document with a finite node-labeled sibling-ordered unranked tree. Let Σ be a finite set of node labels (with a special label $\mathtt{str}$) and C an infinite set of text values. We represent our XML documents with a structure, called *XML Tree*, defined as follows:

Definition 2 (XML Tree). *An XML tree T over Σ is a structure $(N, root, R_\downarrow, R_\rightarrow, \lambda, \nu)$, where N is a set of nodes, $root \in N$ is a distinguished root node, $R_\downarrow \subseteq N \times N$ is the parent-child relation, $R_\rightarrow \subseteq N \times N$ is a successor relation on (ordered) siblings, $\lambda : N \rightarrow \Sigma$ is a function assigning to every node its label, and $\nu : N \rightarrow C$ is a function assigning a text value to each node with label $\mathtt{str}$.*

The relations $R_{\downarrow*}$ and $R_{\rightarrow*}$ represent the reflexive transitive closure of $R_\downarrow$ and $R_\rightarrow$ respectively. We use $R_\uparrow$ and $R_\leftarrow$ to denote the converse of $R_\downarrow$ and $R_\rightarrow$ respectively. In addition, $R_{\uparrow*}$ and $R_{\leftarrow*}$ denote respectively the converse of $R_{\downarrow*}$ and $R_{\rightarrow*}$. Contrary to the model defined in [28], we define the function ν to associate data values with nodes since data value comparison is supported by our XPath fragments defined subsequently.

Definition 3 (Validation of XML trees w.r.t DTD [28]). *An XML tree $T = (N, r, R_\downarrow, R_\rightarrow, \lambda, \nu)$, defined over the set Σ of node labels, conforms to a DTD $D = (Ele, P, root)$ if the following conditions hold:*

1. *The root of T is mapped to root (i.e. $\lambda(r) = root$);*
2. *Each node in T is labeled either with an element type A in Ele, called an A element, or with $\mathtt{str}$, called a text node, therefore $\Sigma = Ele \cup \{\mathtt{str}\}$;*
3. *For each A element with k ordered children $n_1, ..., n_k$, the word $\lambda(n_1), ..., \lambda(n_k)$ belongs to the regular language defined by $P(A)$;*
4. *Each text node n (i.e. with $\lambda(n) = \mathtt{str}$) carries a string value $\nu(n)$ (i.e. PCDATA) and is the leaf of the tree.*

Note that elements of T are a set of nodes of N that are labeled with Ele, while nodes represent both elements and text nodes (i.e. nodes labeled with str). Subsequently, we use the terms of node and element interchangeably.

We call T an instance of a DTD D if T conforms to D. We denote by $T(D)$ the set of all XML trees that conform to D. For instance, Fig. 1 depicts[6] an XML document that conforms to the *department* DTD of Example 1.

2.3   XPath Queries

We define here the different fragments of XPath [23] that are used throughout this paper.

Definition 4 (XPath Downward fragment). *We denote by X the downward fragment of XPath [36] that is defined as follows:*

$$
\begin{aligned}
p &:= \alpha::\eta \mid p\,[q]\cdots[q] \mid p/p \mid p \cup p \\
q &:= p \mid p{=}c \mid q \wedge q \mid q \vee q \mid \neg\,(q) \\
\alpha &:= \varepsilon \mid \downarrow \mid \downarrow^{+} \mid \downarrow^{*}
\end{aligned}
$$

where p denotes an XPath query and it is the start of the production, η is a node test that can be an element type, $*$ (that matches all types), or function text() (that tests whether a node is a text node), c is a string constant, and $\cup, \wedge, \vee, \neg$ denote union, conjunction, disjunction, *and* negation *respectively; α stands for XPath axis relations and can be one of ε, $\downarrow$, $\downarrow^{+}$, or $\downarrow^{*}$ which denote* self, child, descendant, *and* descendant-or-self *axis respectively. Finally the expression q is called a* qualifier, filter *or* predicate.

A qualifier q is said *valid* at a node n, denoted by $n \vDash q$, if and only if one of the following conditions holds: (*i*) q is an atomic predicate that, when evaluated over n, returns at least one node (i.e. there are some nodes reachable from n via q); (*ii*) q is given by $\alpha::text()=c$ and there is at least one node, reachable according to axis α from n, that has a text node with value c; (*iii*) q is a boolean expression and it is evaluated to true at n (e.g. $n \vDash \neg(q)$ if and only if the query q evaluates to empty set over n). We note by $S[\![Q]\!](T)$ the set of nodes resulted from the evaluation of the query Q over the XML tree T.

We define in the following more expressive fragments of XPath that are used to overcome the query rewriting limitation discussed latter in Sect. 3.2.

Definition 5 (Extended fragment). *We consider an extended fragment of X, denoted by $X_{[n,=]}^{\Uparrow}$, and defined as follows:*

$$
\begin{aligned}
p &:= \alpha::\eta \mid p\,[q]\cdots[q] \mid p/p \mid p \cup p \mid p\,[1] \\
q &:= p \mid p{=}c \mid q \wedge q \mid q \vee q \mid \neg\,(q) \mid p = \varepsilon::* \\
\alpha &:= \varepsilon \mid \downarrow \mid \downarrow^{+} \mid \downarrow^{*} \mid \uparrow \mid \uparrow^{+} \mid \uparrow^{*}
\end{aligned}
$$

---

[6] We recall that indices in our examples of XML trees are used to distinguish between elements of the same type, e.g. $course_1$ and $course_2$. Moreover, because of space limitation we focus only on some nodes while $\bigwedge$ denotes the remaining ones.

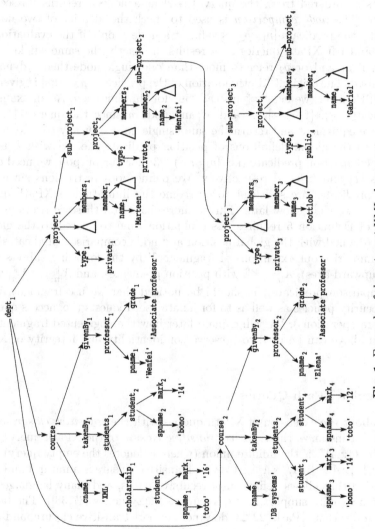

Fig. 1. Example of *department* XML document.

We enrich then $\mathcal{X}$ by the three upward-axes parent *($\uparrow$),* ancestor *($\uparrow^+$), and* ancestor-or-self *($\uparrow^*$), as well as the* position *and the* node comparison *predicates [23].*

In general [23], the position predicate, defined with $[k](k \in \mathbb{N})$, is used to return the kth node from an ordered set of nodes. For instance, since we model XML documents as ordered trees, the query $\downarrow::*[2]$ at a node n returns its second child node. The *node comparison* is used to check the identity of two nodes. Specifically, the predicate $[p_1=p_2]$ is valid at a node n only if the evaluation of the right and left XPath queries at n results in exactly the same single node. Note that if p_1 and/or p_2 refer to more than one single node then a dynamic error is raised. The original XPath notation of the predicate $[p = \varepsilon::*]$ is given by $[p$ **is** $\varepsilon::*]$. However, we use "=" instead "**is**" for simplification. As an example, the predicate $[\uparrow^*::*[1]=\varepsilon::*]$ is valid at any node n since the queries $\uparrow^*::*[1]$ and $\varepsilon::*$ are equivalent and return the same single node over any context node. Contrary to the global definitions of position predicate (i.e. $[k]$ with $k \in \mathbb{N}$) and node comparison predicate (i.e. $[p_1=p_2]$) [23], for our purpose we need only the forms $[1]$ and $[p=\varepsilon::*]$ respectively. We define both restrictions since the resulting predicates are sufficient to overcome the limitation of XPath query rewriting as we shall show later. Furthermore, based on these restrictions our fragment of Definition 5 requires less evaluation time compared to the global fragment (defined with the global position and node comparison predicates).

We summarize our extensions of fragment $\mathcal{X}$ by the following subsets: $\mathcal{X}^{\Uparrow}$ ($\mathcal{X}$ with upward-axes), $\mathcal{X}^{\Uparrow}_{[n]}$ ($\mathcal{X}^{\Uparrow}$ with position predicate), and $\mathcal{X}^{\Uparrow}_{[n,=]}$ ($\mathcal{X}^{\Uparrow}_{[n]}$ with node comparison predicate). It should be noticed that we use fragment $\mathcal{X}$ to specify security policies as well as to formulate user requests (i.e. access queries and update operations). We will explain later how the augmented fragments of $\mathcal{X}$ defined above can be used to preserve confidentiality and integrity of XML data.

2.4 Regular XPath Queries

We talk about the extension of XPath queries with the *transitive closure operator* "*". For instance, the *reflexive transitive closure* of the XPath query $\downarrow::A$, denoted by $(\downarrow::A)^*$, is the infinite union (where ϵ denotes the empty query): $\epsilon \cup \downarrow::A \cup \downarrow::A/\downarrow::A \cup \downarrow::A/\downarrow::A/\downarrow::A \cup \ldots$. Transitive closure is a natural and useful operation that allows definition of recursive paths, and many languages for semistructured data support it (e.g. recursive SQL queries [37,38]). The major concern here is that XPath [22,23] does not support transitive closure, and thus arbitrary recursive paths are not expressible in this language [39].

In spite of its clear practical benefits, no XML engine supports the transitive closure operator. This has led researchers to define some extensions of the XPath language in order to enable definition of recursive path expressions. A useful study is given in [28] to know more about the theoretical properties of XPath 1.0 extended with regular path expressions. Based on their definitions, our class of Regular XPath queries, denoted by $\mathcal{X}_{reg}$, is defined as follows ($p*$ denotes an infinite repetition of the query p):

$$p \; := \; \alpha::ntst \; \mid \; p* \; \mid \; p\,[q]\cdots[q] \; \mid \; p\,/p \; \mid \; p \, \cup \, p$$
$$q \; := \; p \; \mid \; p\texttt{=}c \; \mid \; q \, \wedge \, q \; \mid \; q \, \vee \, q \; \mid \; \neg \, (q)$$
$$\alpha \; := \; \varepsilon \; \mid \; \downarrow \; \mid \; \downarrow^{+} \; \mid \; \downarrow^{*}$$

Based on the formal evaluation algorithm of $\mathcal{X}_{reg}$ queries described in [17], that relies on MFAs (*Mixed Finite state Automatas*), we get the following results:

Proposition 1. *Given an $\mathcal{X}_{reg}$ query Q defined over a DTD D, Q can be translated into an equivalent MFA M of size at most $O(|Q|.|D|)$ in at most $O(|Q|^2.|D|^2)$ time. Moreover, M can be evaluated over any instance T of D in at most $O(|Q|^2.|D|^2 + |Q|.|D|.|T|)$ time and space.*

3   Problem Statement

We present in this section the basic problem we tackle, namely answering XML queries over recursive security views.

3.1   XML Security Views

The notion of *security view*, introduced first by [40], consists on defining for each group of users a view of the underlying XML document that displays all and only parts of the document these users are allowed to access. Fan et al. [25] refined this notion by introducing first a language to specify fine-grained access control policies and a rewriting algorithm to enforce such policies. Security views are now the basic of most existing XML access control models [16–19,21,24,25,27,41,42].

Let T be an XML document that conforms to a DTD D. This document may be queried simultaneously by different users having different access privileges. An access control policy, as defined in [25], is an extension of the document DTD D associating *accessibility conditions* to element types of D. These conditions specify elements of T the users are granted access to. More specifically, an access specification is defined as follows:

Definition 6 (Access Specification [25]). *An access specification S is a pair (D, **ann**) consisting of a DTD D and a partial mapping **ann** such that, for each production rule $A \rightarrow P(A)$ and each element type B in $P(A)$, **ann**(A, B), if explicitly defined, is an annotation of the form:*

$$\textbf{\textit{ann}}(A, B) := Y \mid N \mid [Q]$$

where [Q] is an XPath predicate. The root type of D is annotated Y by default.

Intuitively, the specification values Y, N, and $[Q]$ indicate that the B children of A elements in an instantiation of D are *accessible*, *inaccessible*, or *conditionally accessible* respectively. If **ann**(A, B) is not explicitly defined, then B inherits the accessibility of A. On the other hand, if **ann**(A, B) is explicitly defined then B may *override* the accessibility inherited from A.

Example 2. We consider the *department* DTD of Example 1 and we define some access privileges for professors. Assume that a professor, identified by his name $PNAME, can access to all his courses information except the information denoting whether or not a given student holds a scholarship. The access specification, $S=(dept, \textbf{ann})$, corresponding to these privileges can be specified as follows:

$$\textbf{ann}\,(dept,\,course) = \underbrace{[\downarrow::givenBy/\downarrow::professor/\downarrow::pname = \text{\$PNAME}]}_{Q_1}$$

$$\textbf{ann}\,(students,\,scholarship) = N$$

$$\textbf{ann}\,(scholarship,\,student) = [\uparrow^{+}::course[Q_1]]$$

Here $PNAME is treated as a constant parameter, i.e. when a concrete value, e.g., *Eichten*, is substituted for $PNAME, the specification defines the access control policy for the professor *Eichten*. Observe that $\textbf{ann}\,(course,\,takenBy)$ is not explicitly defined, which means that in an instantiation of the *department* DTD, an *takenBy* element inherits its accessibility from its parent element *course*, this accessibility is either Y or N according to the evaluation of the predicate $[Q_1]$ at this *course* element. Similarly for *cname*, *students*, *givenBy* and his descendant types. The annotation $\textbf{ann}\,(students,\,scholarship)=N$ over a *scholarship* element overrides the accessibility inherited from its ancestor *course* to make this *scholarship* element inaccessible. Moreover, the annotation $\textbf{ann}\,(scholarship,\,student)=[\uparrow^{+}::course[Q_1]]$ overrides the accessibility N, inherited from *scholarship* element, and indicates that *student* children of *scholarship* elements are conditionally accessible (i.e. they are accessible if the professor is granted to access to their ancestor element *course*). □

Access control policies based on the specification of Definition 6 are enforced through the derivation of a *security view* [25]. A security view is an extension of the original XML document and the DTD that: (*1*) may be automatically derived from an access specification, (*2*) displays to the user all and only accessible parts of the XML document, (*3*) provides the user with a schema of all his accessible data so he can formulate and optimize his queries, and (*4*) allows a safe translation of user queries to prevent access to sensitive data[7]. More formally, an XML security view is defined as follows:

Definition 7 (Security View [25]). *Given an access specification $S=(D, \textbf{ann})$ defined over a non-recursive DTD D, a security view V is a pair (D_v, σ) where D_v is the DTD view of D that presents the schema of all and only data the user is granted access to, and σ is a function defined as follows: for any element type A and its child type B in D_v, $\sigma(A, B)$ is a set of XPath expressions that when evaluated over an A element of an XML tree T of D, returns all its accessible children B. In other words, σ maps each instance of D to an instance of D_v that contains only accessible data.*

---

[7] This translation is necessary only if the views of the data are virtual, i.e. not materialized.

The DTD view D_v is given to the user for formulation and optimization of queries. However, the set of XPath expressions defined by σ are hidden from the user and used to extract for any XML tree $T \in \mathcal{T}(D)$, a view T_v of T that contains all and only accessible nodes of T.

Example 3. Consider the access specification $S=(dept, \textbf{ann})$ of Example 2. The DTD view $dept_v=(\Sigma_v, dept, P_v)$ of the *department* DTD can be computed by eliminating the *scholarship* element type, i.e. $\Sigma_v := \Sigma \setminus \{scholarship\}$, and by changing the definition of *dept* and *students* element types as follows:

$$P_v(dept) \quad := (course*, project*)$$
$$P_v(students) := (student+)$$
$$P_v(A) \quad := P(A), \text{ for all remaining element types } A \text{ in } \Sigma_v$$

The function σ is defined over the production rules of $dept_v$ as follows: (refer to Example 2 for the definition of $[Q_1]$)

$dept \quad \rightarrow P_v(dept):$
$\qquad \sigma(dept, course) = \downarrow::course[Q_1]$
$\qquad \sigma(dept, project) = \downarrow::project$

$students \rightarrow P_v(students):$
$\qquad \sigma(students, student) =$
$\qquad \downarrow::student \cup \downarrow::scholarship/\downarrow::student[\uparrow^+::course[Q_1]]$

$A \qquad \rightarrow P_v(A):$ (for each remaining element type A in Σ_v)
$\qquad \sigma(A, B) = \downarrow::B$ (for each child type B in $P_v(A)$)

Using the resulting security view $V=(dept_v, \sigma)$, the view of the XML document of Fig. 1 is derived and depicted in Fig. 2, this view shows all and only parts of the original XML document that are accessible w.r.t the specification $S=(dept, \textbf{ann})$. Note that all descendants of the element $project_1$ are still unchanged. $\square$

Given a security view $V=(D_v, \sigma)$ defined for an access specification $S=(D, \textbf{ann})$, then, for each instance T of D and its view T_v computed using the σ function, one can either materialize T_v and evaluate user queries directly over it [24,43], or keep T_v virtual for some reasons [17–19,21]. In case of virtual views, the *query rewriting* principle is used to translate each user query Q defined in D_v over the virtual view T_v, into a safe one Q^t defined in D over the original document T such that: evaluating Q over T_v returns the same set of nodes as the evaluation of the rewritten query Q^t over T.

Example 4. Consider the query $\downarrow::dept/\downarrow::course$ of the professor $Wenfei$ defined over the view of Fig. 2. This query can be rewritten, using the security view of Example 3, to $\downarrow::dept/\sigma(dept, course)$ that is equal to:

$$\downarrow::dept/\downarrow::course[\downarrow::givenBy/\downarrow::professor/\downarrow::pname=\text{``}Wenfei\text{''}]$$

The evaluation of this query over the original XML document of Fig. 1 returns only accessible *course* elements, i.e. $course_1$. $\square$

Since most existing approaches for securing XML data are based on the security view model, we discuss thereafter the major limits of this model.

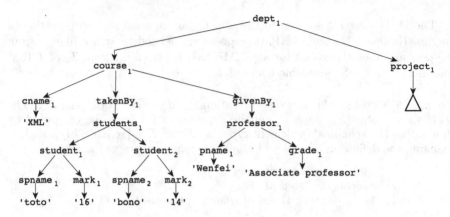

Fig. 2. The view of the *dept* XML document w.r.t the policy of Example 2.

3.2 Security View's Drawbacks

We study only the case of querying virtual XML data, then problems related to manipulation of materialized XML views [24,43] are outside the topic of interest of this work. More precisely, we discuss subsequently obstacles encountered when manipulating recursive views[8] and this at the stage of query rewriting. Even if the rewriting of XPath queries is quite straightforward for non-recursive XML security views, some obstacles may arise in the presence of recursive views that make this rewriting process impossible for some class of XPath queries. More precisely, the rewriting process is based on the definition of the function σ that, in case of recursive DTDs, cannot be defined in XPath as we show by the following example.

Example 5. We consider the *department* DTD of Example 1 and we assume that a personal of some department, identified by his name $PNAME, can access to information of any project in which he is a member, as well as information of all public projects. The access specification, $S=(dept, \boldsymbol{ann})$, corresponding to these privileges is defined with:

$$\underbrace{\boldsymbol{ann}(dept, project) = \boldsymbol{ann}(sub\text{-}project, project) =}_{Q_2}$$
$$\underbrace{[\downarrow::type/\downarrow::public \vee \downarrow::members/\downarrow::member[\downarrow::name = \$\text{PNAME}]]}_{Q_2}$$

Note that if the predicate $[Q_2]$ is valid at a given *project* element then all its descendant elements inherit this accessibility except *sub-project* elements that may override it (that depends to the evaluation of $[Q_2]$). Consider the case of the professor *"Wenfei"*, the view of the XML document of Fig. 1 is derived and depicted in Fig. 3. Given an accessible *dept* element, there is an infinite set of paths that connect this element to its accessible children of type *project*. More precisely, $\sigma(dept, project)$ can be defined using the transitive closure operator "*" with: $\sigma(dept, project) = (\downarrow::project[\neg(Q_2)]/\downarrow::sub\text{-}project)*/\downarrow::project[Q_2]$.

---

[8] A security view is *recursive* if it is defined over a recursive DTD.

The recursive path $(\downarrow::project[\neg(Q_2)]/\downarrow::sub\text{-}project)*$ is defined over only inaccessible elements. Thus, the expression $\sigma(dept, project)$ has to extract, over each accessible element of type $dept$ in the original data, the accessible descendants of type $project$ that appear in the view of the data as immediate children of this $dept$ element. In other words, an element m of type $project$ is shown in the view of Fig. 3 as an immediate child of some $dept$ element n if and only: m and n are both accessible in the original tree of Fig. 1, and either m is an immediate child of n or separated from n with only inaccessible elements. Take the case of the elements $dept_1$ and $project_2$ of the tree of Fig. 1. After hiding the inaccessible element $project_1$, $project_2$ appears in the view of Fig. 3 as immediate child of $dept_1$. The same principle is applied for the elements $project_2$ and $project_4$. □

Authors of [28] showed that the kleen star operator "*" cannot be expressed in XPath. For this reason, the function σ of Example 5 cannot be defined in the standard XPath which makes the query rewriting process more challenging. We are principally motivated by studding the *closure* of a significant class of XPath queries (denoted by $\mathcal{X}$) under query rewriting, i.e. whether all queries of this class can be rewritten over arbitrary security views (recursive or not). We define formally the *closure property* as follows:

Definition 8. *An XML query language L is* closed *under query rewriting if there exists a function* $\mathcal{R}: L \to L$ *that, for any access specification* $S = (D, ann)$ *and any DTD view* D_v *of D, translates each query Q of L defined over* D_v *into another one* $\mathcal{R}(Q)$ *defined in L over D such that: for any* $T \in \mathcal{T}(D)$ *and its virtual view* T_v, $\mathcal{S}[\![\mathcal{R}(Q)]\!](T) = \mathcal{S}[\![Q]\!](T_v)$.

Note that Fan et al. [25] shown that the fragment $\mathcal{X}$ (Definition 4) is closed under query rewriting in case of non-recursive security views. However, in case of recursion, that is no longer the case as shows the following theorem:

Theorem 1 ([17,36]). *In case of recursive XML security views, the XPath fragment* $\mathcal{X}$ *is not closed under query rewriting.*

Finally, we should emphasize that no practical solution exists to respond to XML queries over recursive security views. Some theoretical results exist that are based on Regular XPath language which allows definition of recursive queries. According to [17,18], the fragment $\mathcal{X}_{reg}$ of Sect. 2.4 is closed under query rewriting. However, some major drawbacks are to be noted: no standard solution exists to evaluate regular queries, Regular XPath evaluation is more costly than standard XPath in general, and since contemporary database systems provide support for XPath only as XML query language, the results found around Regular XPath are still impractical.

4   Access Control with Arbitrary DTDs

Figure 4 presents our XML access control framework. It is designed particularly for native XML databases where XML data is stored in its native format. The

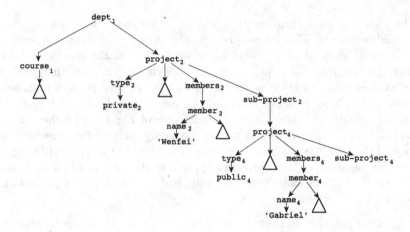

Fig. 3. The view of the *dept* XML document w.r.t the policy of Example 5.

module *Policy Specifier* allows the administrator to specify, for each group of users, the document they can query and an access control policy to handle this querying. According to this policy, the module *View Generator* computes a virtual view of their related document as well as a view (or an approximated view) of its corresponding DTD. This DTD view is used by the users to formulate their queries and query the virtual data view that is provided to them. Recall that the fragment $\mathcal{X}$ is used for user queries formulation. Each $\mathcal{X}$ query is rewritten into a safe one, defined in the fragment $\mathcal{X}_{[n,=]}^{\Uparrow}$, and evaluated over the original document. The results of this evaluation are given to the user as a set of sub-trees where each one presents an accessible node referred to by the input query.

We present in the following the *hospital DTD* that corresponds to a real-life patient medical data [44] and which is used throughout the rest of this paper.

Example 6. The *hospital DTD* $(\Sigma, P, hospital)$ is defined with the following production rules (definitions of elements whose type is **str** are omitted):

$$
\begin{aligned}
hospital &\rightarrow (department^*)\\
department &\rightarrow (name, patient^*)\\
patient &\rightarrow (pname, wardNo, parent?, sibling?,\\
&\quad\ symptoms^*, intervention)\\
parent &\rightarrow (patient^*)\\
sibling &\rightarrow (patient^*)\\
symptoms &\rightarrow (symptom^*)\\
intervention &\rightarrow (doctor, treatment)\\
doctor &\rightarrow (dname, specialty)\\
\\
treatment &\rightarrow (type, Tresult^*, diagnosis)\\
diagnosis &\rightarrow (Dresult^*, implies?)\\
implies &\rightarrow (treatment \mid intervention)
\end{aligned}
$$

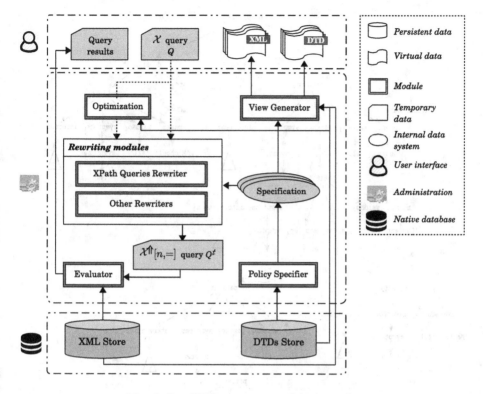

Fig. 4. Our XML access control framework.

A hospital DTD document consists of a list of departments, each *department* (defined by its *name*) has a list of patients currently residing in the hospital. For each *patient*, the hospital maintains her name (*pname*), a ward number (*wardNo*), a family medical history by means of the recursively defined *parent* and *sibling* relations, as well as a list of *symptoms*. The hospitalization is marked by the *intervention* of one or many doctors depending on their specialty and the patient care requirement. For each intervention, the hospital also maintains the responsible *doctor* (represented by its name *dname* and *specialty*), and the *treatment* applied. A treatment is described by its *type*, a list of result (*Tresult*), and it is followed by a *diagnosis* phase. According to the diagnosis results (*Dresult*), either another treatment is planned or the intervention of another doctor/specialist/expert is solicited[9]. An instance of this hospital DTD is given in Fig. 5 (some text contents are abbreviated by '...').   □

---

[9] According to [44], this may happen when the required treatment is outside the area of expertise of the current responsible doctor.

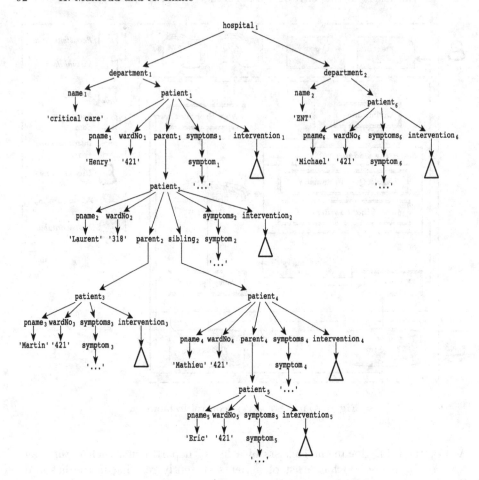

Fig. 5. Example of hospital data.

4.1 Access Specification

Fan et al. [25] proposed the first language for the specification of XML access control policies through annotation of DTD grammars. Moreover, authors of [24] studied the classification of such policies w.r.t the default annotation, the inheritance and the overriding of annotations. In this work we consider only the case of *top-down* access control policies where the root node of the XML tree is accessible by default and each intermediate node can either inherit the annotation of its parent node or override it (see Definition 6). Although both access specification languages defined in [24, 25] are based on the same principle, i.e. annotating element types of DTDs with Y, N and $[Q]$, there is a significant difference in the use of conditional annotations (i.e. annotations of the form $[Q]$). We consider the following example for more details:

Example 7. We suppose that there are two annotations $ann(A, B)=[\neg (\downarrow::D)]$ and $ann(C, D)=Y$ defined over a simple XML tree composed by only one path:

$$R \rightarrow A \rightarrow B \rightarrow C \rightarrow D$$

Note that the predicate $[\neg (\downarrow::D)]$ is invalid at the element node B. According to [25], all the subtree rooted at this B element is inaccessible and thus the second annotation that concerns the element node D does not take effect. According to [24] however, the element node D overrides the value N inherited from its ancestor element B and becomes accessible.   □

In general, let n be an element node that is concerned by an annotation of the form $[Q]$. For the former work, if $n \nvDash Q$ then all the subtree rooted at n is inaccessible and no annotation defined over descendants of n can take effect. For the second work however, even if $n \nvDash Q$, descendants of n can override this annotation to become accessible.

We assume that the two definitions are useful and in practice applications may require the application of both kinds of annotations, even within the same scenario. For this reason, we present a refined and more expressive access specification language whose access specifications are defined as follows:

Definition 9 (Extended Access Specification). *We define an access specification S as a pair (D, ann) consisting of a DTD D and a partial mapping ann such that, for each production rule $A \rightarrow P(A)$ and each element type B in $P(A)$, $ann(A, B)$, if explicitly defined, is an annotation of the form:*

$$ann(A, B) := Y \mid N \mid [Q] \mid N_h \mid [Q]_h$$

where $[Q]$ is an XPath predicate. Annotations of the form N_h and $[Q]_h$ are called downward-closed annotations. *The root type of D is annotated Y by default.*

Recall from Definition 6 that annotations of the form Y, N, and $[Q]$ indicate that an B element, child of an A element, is *accessible*, *inaccessible*, or *conditionally accessible* respectively. We allow overriding between annotations of the three previous forms. In other words, each element concerned by an annotation of the form Y, N, or $[Q]$ overrides its inherited annotation if it is defined with one of these three forms. The special specification values N_h and $[Q]_h$ indicate that overriding is *denied* or *conditionally allowed* respectively. More specifically, let $n_1, \ldots, n_l$ $(l \geq 2)$ be element nodes of types $A_1, \ldots, A_l$ respectively where each n_i $(1 \leq i < l)$ is parent node of n_{i+1}. The annotation $ann(A_1, A_2)=N_h$ indicates that all the subtree rooted at n_2 is inaccessible and no element under n_2 can override this annotation. Thus, if some annotation $ann(A_i, A_{i+1})=Y \mid [Q]$ is explicitly defined then the element node n_{i+1} remains inaccessible even if $n_{i+1} \vDash Q$. However, the annotation $ann(A_1, A_2)=[Q_2]_h$ indicates that annotations defined over descendant types of A_2 take effect only if Q_2 is valid. In other words, given the annotation $ann(A_i, A_{i+1})=Y$ (resp. $[Q_{i+1}]$), the element node n_{i+1} is accessible if and only if: $n_2 \vDash Q_2$ (resp. $n_2 \vDash Q_2 \wedge n_{i+1} \vDash Q_{i+1}$).

Example 8. Suppose that the hospital wants to impose some restrictions that allow some nurse to access only information of patients who are being treated in the *critical care* department and residing at the ward *421*. In addition, all sibling data should be inaccessible. This policy can be specified using our specification language with an access specification $S=(D, ann)$ where D is the hospital DTD and the function ann defines the three following annotations:

R_1: $ann\,(hospital,\ department)=\underbrace{[\downarrow ::name = \text{``critical care''}}_{Q_1}]_h$

R_2: $ann\,(department,\ patient)=ann\,(parent,\ patient)=\underbrace{[\downarrow ::wardNo = \text{``421''}}_{Q_2}]$

R_3: $ann\,(patient,\ sibling)=N_h$

According to this specification, the view of the data of Fig. 5 is extracted and depicted in Fig. 6. This view displays all and only the data the nurse is granted access to. All the data of the *ENT* department is hidden, i.e. the subtree rooted at the $departement_2$ element. Since R_1 is downward-closed and $departement_2 \not\models Q_1$, then the annotation R_2 cannot be applied at $patient_6$ element which remains inaccessible even with $patient_6 \models Q_2$. Notice that $departement_1 \models Q_1$ which means that the $departement_1$ element is accessible and overriding of annotations is allowed for its descendants. Thus, the elements $patient_1$ and $patient_3$ are accessible along with their immediate children since Q_2 is valid at these elements, while the element $patient_2$ (with $patient_2 \not\models Q_2$) overrides the annotation Y inherited from $patient_1$ and becomes inaccessible along with all its immediate children. In this way, $patient_3$ element appears at the view of Fig. 6 as immediate child of $parent_1$. Finally, since $sibling_2$ element is concerned by the downward-closed annotation R_3 with value N_h, then all the subtree rooted at $sibling_2$ is inaccessible and annotation R_2 cannot take effect over the elements $patient_4$ and $patient_5$.   □

Our access specification language is more expressive than existing ones in the sens that the access policies of many current approaches can be specified in our language using only few annotation values as shown in Table 1. For instance, the policy of Example 8 cannot be specified in the fragment $\mathcal{X}$ using the specification languages presented in [24, 25]. This can be done using a more expressive fragment, like $\mathcal{X}^{\Uparrow}$, but the annotations may be more verbose and difficult to manage.

The *completeness* and *consistency* of access control policies have been defined in [45] as follows. Let P be an access control policy and T be an XML tree. If a node n in T is not concerned by any access rule of P then P is *incomplete*. Moreover, if there are both a negative and a positive access rule for the same node n (i.e. n is both accessible and inaccessible) then P is *inconsistent*. Consider our access specifications of Definition 9, we define the notions of *completeness* and *consistency*, along the same lines as [24, 25], as follows:

Definition 10. *Given an access specification $S=(D, ann)$ and an XML tree $T \in \mathcal{T}(D)$, then, we say that S is* complete *and* consistent *if and only if the*

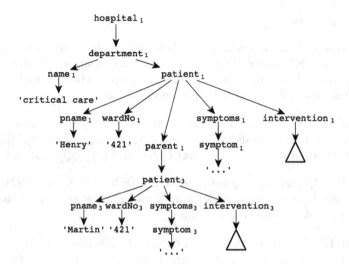

Fig. 6. View of the tree of Fig. 5 computed w.r.t the policy of Example 8.

accessibility *of each node in T is* uniquely *defined, i.e. it is either* accessible *or* inaccessible.

Proposition 2. *The access control policies based on Definition 9 are complete and* consistent.

Proof. Authors of [24] have proved that access policies defined with specification values of the form Y, N and $[Q]$ are complete and consistent. The case of downward-closed annotations is straightforward and the proof of the latter work can be easily extended to handle this kind of annotations.   □

Table 1. Current approaches' policies specified with our language.

Access policies	Required specification values					Remark
	Y	N	N_h	$[Q]$	$[Q]_h$	
[17, 25, 27]	✓	✓			✓	
[24]	✓	✓		✓		case of top-down policies
[18]	✓	✓		✓		
[21]	✓		✓	✓		
[19]	✓	✓				
[46]	✓		✓	✓		*deny overwrites* as the conflict resolution policy
[47]	✓		✓	✓		with *denial downwards* consistency requirement

4.2   Accessibility

The enforcement of our access control policies relies principally on the definition of *node accessibility*. Inspired from [18,46], we define a single XPath filter, that can be constructed for any access specification, which checks whether a given XML node is *accessible* or not w.r.t this specification.

Definition 11. *Let n be an B element that is child of an A element. A given annotation* $ann(A, B)$ *is* valid *at n if and only if* $ann(A, B)=Y\,|[Q]|[Q]_h$ *with* $n \vDash Q$. *Otherwise, it is* invalid, *i.e.* $ann(A, B)=N\,|N_h|[Q]|[Q]_h$ *with* $n \nvDash Q$.

If $ann(A, B)=[Q]_h$ with $n \vDash Q$ (resp. $ann(A, B)=N_h|[Q]_h$ with $n \nvDash Q$) then we talk about *valid* (resp. *invalid*) downward-closed annotation. Given the above, we define the node accessibility as follows:

Definition 12. *Let* $S=(D, ann)$ *be an access specification, T be an instance of D, and n be an element node in T of type B having parent node of type A. The element node n is* accessible *w.r.t S if and only if the following conditions hold:*

(i) Either there exists an explicitly defined annotation $ann(A, B)$ *that is valid at n; or the first annotation explicitly defined over ancestors of n is valid.*
(ii) There is no invalid downward-closed annotation defined over ancestors of n.

More specifically, consider the element nodes $n_1, \ldots, n_k$ ($k \geq 2$) of element types $A_1, \ldots, A_k$ respectively where n_1 is the root node. Take the case of the element node n_k, the condition (i) of Definition 12 refers to one of the following three cases:

(a) Only the default annotation $ann\ (A_1)=Y$ is defined over the types $A_1, \ldots, A_k$. Thus, n_k inherits its accessibility from the root node n_1.
(b) The annotation $ann\ (A_{k-1}, A_k)$ is explicitly defined and valid at n_k.
(c) The annotation $ann\ (A_{i-1}, A_i)$ is explicitly defined and valid at the element n_i ($1 < i < k$), and no annotation is defined over the types $A_{i+1}, \ldots, A_k$. Thus, n_k inherits its accessibility from its ancestor node n_i.

The condition (ii) of Definition 12 implies that for any downward-closed annotation $ann\ (A_{i-1}, A_i)$ defined over ancestor n_i of n_k (with $1 < i < k$), either $ann\ (A_{i-1}, A_i) \neq N_h$ or $ann\ (A_{i-1}, A_i)=[Q]_h$ with $n_i \vDash Q$. Finally, note that a text node is accessible if and only if its parent element is accessible.

Definition 13. *Given an access specification* $S=(D, ann)$, *we define two* $\mathcal{X}_{[n]}^{\Uparrow}$ *predicates* $\mathcal{A}_1^{acc}$ *and* $\mathcal{A}_2^{acc}$ *as follows:*

$\mathcal{A}_1^{acc} := \uparrow^*::*[allAnn]/[1][validAnn]$, *where:*
$allAnn := \varepsilon::root \bigvee_{ann(A',A)\in ann} \varepsilon::A/\uparrow::A'$
$validAnn := \varepsilon::root \bigvee_{(ann(A',A)=Y)\in ann} \varepsilon::A/\uparrow::A' \bigvee_{(ann(A',A)=[Q]|[Q]_h)\in ann}$
$\varepsilon::A[Q]/\uparrow::A'$

$$\mathcal{A}_2^{acc} := \bigwedge_{(ann(A',A)=[Q]_h)\in ann} \neg\ (\uparrow^+::A[\neg\ (Q)]/\uparrow::A')\ \bigwedge_{(ann(A',A)=N_h)\in ann} \neg$$
$$(\uparrow^+::A/\uparrow::A')$$

The predicates $\mathcal{A}_1^{acc}$ and $\mathcal{A}_2^{acc}$ satisfy the conditions (i) and (ii) of Definition 12 respectively.

The first predicate checks whether the node n is explicitly concerned by a valid annotation (case **b**) or inherits its accessibility from a valid annotation defined over its ancestors (cases **a** and **c**). The second predicate checks whether the node n is not in the scope of an invalid downward-closed annotation. The predicate [allAnn] consists of a disjunction of all annotations, while [validAnn] presents disjunction of only valid annotations. More precisely, the evaluation of the predicate $\uparrow^*::*[allAnn]$ at a node n returns an ordered set of nodes N that contains the node n and/or some of its ancestors such that each one is "explicitly" concerned by an annotation of S, i.e. $N \subseteq \{n\} \cup \boldsymbol{ancestors}(n)^{10}$, and $\forall m \in N$, m is of type B and has a parent node of type A where $ann\,(A,B)$ is explicitly defined in S. The predicate $\uparrow^*::*[allAnn][1]$ (i.e. $N[1]$) returns the first node in N, i.e. either the node n (if it is explicitly concerned by an annotation), the first ancestor of n that is explicitly concerned by an annotation, or the root node (if only the default annotation is defined). The last predicate [validAnn] checks whether the annotation defined over the node $N[1]$ is valid: this means that either the node n is explicitly concerned by a valid annotation or it inherits its accessibility from one of its ancestors that is concerned by a valid annotation (condition (i)). The use of the second predicate $\mathcal{A}_2^{acc}$ is obvious: if $n \vDash \mathcal{A}_2^{acc}$ then all the downward-closed annotations defined over $\boldsymbol{ancestors}\,(n)$ are valid (condition (ii)).

Lemma 1. *Given an access specification $S=(D, \boldsymbol{ann})$, we define the accessibility predicate $\mathcal{A}^{acc}:=\mathcal{A}_1^{acc} \wedge \mathcal{A}_2^{acc}$ such that: for any XML tree $T \in \mathcal{T}(D)$, a node n of T is accessible if and only if $n \vDash \mathcal{A}^{acc}$.*

According to this lemma, for any access specification $S=(D, \boldsymbol{ann})$ and any XML tree $T \in \mathcal{T}(D)$, the query $\downarrow^*::*[\mathcal{A}^{acc}]$ over T returns the set of all accessible nodes of T where $\mathcal{A}^{acc}$ is computed w.r.t S.

Example 9. Consider the access policy of nurses defined in Example 8 with the following annotations:

$$ann\,(hospital, department)=[\underbrace{\downarrow::name = \text{``critical care''}}_{Q_1}]_h$$

$$ann\,(department, patient)=ann\,(parent, patient)=[\underbrace{\downarrow::wardNo = \text{``421''}}_{Q_2}]$$

$$ann\,(patient, sibling)=N_h$$

According to these annotations, the predicates $\mathcal{A}_1^{acc}$ and $\mathcal{A}_2^{acc}$, that compose $\mathcal{A}^{acc}$, are defined as follows:

---

[10] We use $\boldsymbol{ancestors}\,(n)$ to refer to all ancestors of the node n.

$\mathcal{A}_1^{acc} := \uparrow^*::*[allAnn][1][validAnn]$, where:
$allAnn := \varepsilon::root \vee \varepsilon::department/\uparrow::hospital \vee \varepsilon::patient/\uparrow::department \vee$
$\varepsilon::patient/\uparrow::parent \vee \varepsilon::sibling/\uparrow::patient$

$validAnn := \varepsilon::department[Q_1]/\uparrow::hospital \vee \varepsilon::patient[Q_2]/\uparrow::department$
$\vee \varepsilon::patient[Q_2]/\uparrow::parent \vee \varepsilon::root$

$\mathcal{A}_2^{acc} := \neg (\uparrow^+::departement[\neg (Q_1)]/\uparrow::hospital) \wedge$
$\qquad \neg (\uparrow^+::sibling/\uparrow::patient)$

Consider the case of the element $patient_1$ of Fig. 5. The predicate $\uparrow^*::*[allAnn]$
at $patient_1$ returns the set $N=\{patient_1, departement_1, hospital_1\}$ (each element is concerned by an explicit annotation). We have $N[1]= \{patient_1\}$ and
the predicate $[validAnn]$ is valid at $patient_1$ (since $patient_1 \vDash Q_2$). Thus, the
predicate $\mathcal{A}_1^{acc}$ is valid at $patient_1$. It is clear to see that $\mathcal{A}_2^{acc}$ is also valid at
$patient_1$. We conclude that $patient_1 \vDash (\mathcal{A}_1^{acc} \wedge \mathcal{A}_2^{acc})$ which means that the element $patient_1$ is accessible. Consider now the element $patient_2$, $\uparrow^*::*[allAnn]$
at $patient_2$ returns the set $N'=\{patient_2, patient_1, departement_1, hospital_1\}$,
$N'[1]=\{patient_2\}$, however, the predicate $[validAnn]$ is not valid at $patient_2$
(since $patient_2 \nvDash Q_2$). Thus, $patient_2 \nvDash \mathcal{A}_1^{acc}$ and then the element $patient_2$ is
not accessible. For the element $patient_4$, although $patient_4 \vDash \mathcal{A}_1^{acc}$, $patient_4$ is
inaccessible since $patient_4 \nvDash \mathcal{A}_2^{acc}$ (i.e. $patient_4$ is descendant of $sibling_2$ element
that is concerned by an invalid downward-closed annotation). Finally, the query
$\downarrow^*::*[\mathcal{A}^{acc}]$ over the Fig. 5 returns all the accessible elements that compose the
view of Fig. 6.   □

5   Query Rewriting

We discuss in this section the basic principle of our XML access control approach. We recall that the fragment $\mathcal{X}$ (see Definition 4) is used in our approach for
specification of access control policies as well as for formulation of user queries.
However, we use more larger fragments of XPath to overcome the query answering problem presented in Sect. 3.2. More precisely, the access control policies
based on Definition 9 are enforced through a rewriting technique. Let $S=(D,$
$ann)$ be an access specification, T be an instance of D, T_v be the virtual view
of T computed w.r.t S, and Q be a query defined in $\mathcal{X}$. Our goal is to define a
rewriting function $Rewrite$ such that:

$$\mathcal{X} \longrightarrow \mathcal{X}_{[n,=]}^{\Uparrow}$$
$$Q \longmapsto Rewrite(Q) \text{ such that } S[\![Rewrite(Q)]\!](T)=S[\![Q]\!](T_v)$$

5.1   Queries Without Predicates

Let us now consider queries without predicates, postponing rewriting of predicates to the next subsection. We consider the case of $\mathcal{X}$ queries of the form
$\alpha_1::\eta_1/\cdots/\alpha_k::\eta_k$ $(k \geq 1)$ where $\alpha_i \in \{\varepsilon, \downarrow, \downarrow^*, \downarrow^+\}$ and η_i can be any element

type, *-label, or *text()* function. The union of queries is discussed later. We show first that the rewriting limitation for this kind of queries is encountered when manipulating the $\downarrow$ axis, however, the remaining axes can be rewritten in a simple manner using only the accessibility predicate.

Example 10. Consider the XML tree of Fig. 5 and its view depicted in Fig. 6 that is computed w.r.t the access policy of Example 8. We suppose the the nurse formulates the query $\downarrow^+::departement/\downarrow^+::patient$ over its data view which returns the nodes $patient_1$ and $patient_3$. It is easy to see that this query can be rewritten over the original data into $\downarrow^+::departement[\mathcal{A}^{acc}]/\downarrow^+::patient[\mathcal{A}^{acc}]$ where the predicate $\mathcal{A}^{acc}$ is given in Example 9. Obviously, this rewritten query selects first accessible *departement* elements of Fig. 5, i.e. $departement_1$ element, and then returns all its accessible descendants of type *patient*, i.e. $patient_1$ and $patient_3$. The accessibility of these nodes are checked using $\mathcal{A}^{acc}$. Consider now another query over the data view of nurses defined by $\downarrow^*::parent/\downarrow::*$ and which must return only the node $patient_3$. Since there is a cycle between the *patient* and *parent* elements of the hospital DTD, this latter query cannot be rewritten using only the accessibility predicate. More precisely, the query $\downarrow^*::parent[\mathcal{A}^{acc}]/\downarrow::*[\mathcal{A}^{acc}]$ over the original document returns no element since it selects first the accessible element $parent_1$, while its immediate child $patient_2$ is not accessible. Moreover, a cycle cannot be captured by replacing $\downarrow$ axes with $\downarrow^*$ axes. The query $\downarrow^*::parent[\mathcal{A}^{acc}]/\downarrow::*[\mathcal{A}^{acc}]$ over the original document returns both the node $patient_3$ as well as other additional elements: $pname_3$, $symptoms_3$, $symptom_3$, etc.   □

We show in the following how that the upward axes and the position predicate of the XPath fragment $\mathcal{X}_{[n]}^{\Uparrow}$ can be used to overcome the rewriting limitation encountered when considering $\mathcal{X}$ queries without predicates.

Definition 14. *Given an access specification $S=(D, \mathbf{ann})$ and an element type B, then we define two $\mathcal{X}_{[n]}^{\Uparrow}$ predicates $\mathcal{A}^+$ and $\mathcal{A}^B$ with: $\mathcal{A}^+ := \uparrow^+::*[\mathcal{A}^{acc}]$, and $\mathcal{A}^B := \uparrow^+::*[\mathcal{A}^{acc}][1]/\varepsilon::B$. For any element node n, the evaluation $S[\![\mathcal{A}^+]\!](\{n\})$ returns all the accessible ancestors of n, while $S[\![\mathcal{A}^B]\!](\{n\})$ returns the first accessible ancestor of n whose type is B.*

Finally, we give the details of our rewriting function. Given an access specification $S=(D, \mathbf{ann})$, we define the function $\mathbf{Rewrite}: \mathcal{X} \longrightarrow \mathcal{X}_{[n]}^{\Uparrow}$ that rewrites any $\mathcal{X}$ query Q, of the form $\alpha_1::\eta_1/\cdots/\alpha_k::\eta_k$ $(k \geq 1)$, into another one defined in the fragment $\mathcal{X}_{[n]}^{\Uparrow}$ as follows:

$$\mathbf{Rewrite}(Q) := \downarrow^*::\eta_n[\mathcal{A}^{acc}][prefix^{-1}(\alpha_1::\eta_1/\cdots/\alpha_k::\eta_k)]$$

The qualifier $prefix^{-1}(\alpha_1::\eta_1/\cdots/\alpha_k::\eta_k)$ presents a recursive rewriting in a descendant manner where each sub-query $\alpha_i::\eta_i$ is rewritten over all the sub-queries that precede it in the query Q. In other words, for each sub-query $\alpha_i::\eta_i$ $(1 \leq i \leq k)$, $prefix^{-1}(\alpha_1::\eta_1/\cdots/\alpha_{i-1}::\eta_{i-1})$ is already computed and used to compute $prefix^{-1}(\alpha_1::\eta_1/\cdots/\alpha_i::\eta_i)$ as follows:[11]

---

[11] For $\alpha_i \in \{\downarrow^+, \downarrow^*\}$, $\alpha_i^{-1}=\uparrow^+$ if $\alpha_i=\downarrow^+$ and $\uparrow^*$ otherwise.

$-\ \alpha_i = \downarrow$:

$$prefix^{-1}(\alpha_1::\eta_1/\cdots/\alpha_i::\eta_i) := \mathcal{A}^{\eta_{i-1}}[prefix^{-1}(\alpha_1::\eta_1/\cdots/\alpha_{i-1}::\eta_{i-1})]$$

$-\ \alpha_i \in \{\downarrow^+, \downarrow^*\}$:

$$prefix^{-1}(\alpha_1::\eta_1/\cdots/\alpha_i::\eta_i) := \alpha_i^{-1}::\eta_{i-1}[\mathcal{A}^{acc}][prefix^{-1}(\alpha_1::\eta_1/\cdots/\alpha_{i-1}::\eta_{i-1})]$$

$-\ \alpha_i = \varepsilon$:

$$prefix^{-1}(\alpha_1::\eta_1/\cdots/\alpha_i::\eta_i) := \varepsilon::\eta_{i-1}[prefix^{-1}(\alpha_1::\eta_1/\cdots/\alpha_{i-1}::\eta_{i-1})]$$

As a special case, the first sub-query is rewritten over the root type. Thus, we have $prefix^{-1}(\downarrow::\eta_1)=\mathcal{A}^{root}$, $prefix^{-1}(\downarrow^+::\eta_1)=\uparrow^+::root$, while for the remaining axes, $\alpha_1 \in \{\varepsilon, \downarrow^*\}$, $prefix^{-1}(\alpha_1::\eta_1)$ is empty.

Example 11. Let us consider the query $Q=\downarrow^*::parent/\downarrow::*$ of Example 10 posed over the data view of Fig. 6. By considering the access specification of Example 8, this query can be rewritten as follows: $Rewrite(Q)=\downarrow^*::[\mathcal{A}^{acc}][\mathcal{A}^{parent}]$. By replacing $\mathcal{A}^{parent}$ with its value, we obtain: $\downarrow^*::[\mathcal{A}^{acc}][\uparrow^+::*[\mathcal{A}^{acc}][1]/\varepsilon::parent]$. Recall that the definition of the predicate $\mathcal{A}^{acc}$ w.r.t the access specification of Example 8 is given in Example 9. The evaluation of the query $\downarrow^*::[\mathcal{A}^{acc}]$ over the original document of Fig. 5 returns a node set N composed by all the accessible nodes depicted in Fig. 6. The evaluation of $[\mathcal{A}^{parent}]$ over the set N returns only those elements having as the first accessible ancestor, an element of type *parent*, thus the query $\downarrow^*::*[\mathcal{A}^{acc}][\mathcal{A}^{parent}]$ over the original document returns the element $patient_3$ that is the only element that satisfies the predicate $[\mathcal{A}^{parent}]$: $\mathcal{S}[\![\mathcal{A}^{parent}]\!](\{patient_3\})$ returns the element $parent_1$, i.e. $patient_3 \models \mathcal{A}^{parent}$. Therefore, the query $Rewrite(Q)$ over the original document of Fig. 5 returns only the element $patient_3$ as does the query Q over the data view of Fig. 6.   □

5.2   Rewriting Predicates

We discuss in this section the rewriting of predicates of the fragment $\mathcal{X}$ to complete the description of our rewriting approach. Given an access specification $S=(D, \boldsymbol{ann})$, we define the function $RW\_Pred: \mathcal{X} \to \mathcal{X}^{\Uparrow}_{[n,=]}$ that rewrites any $\mathcal{X}$ predicate P, of the form $\alpha_1::\eta_1/\cdots/\alpha_k::\eta_k$ $(k \geq 1)$, into another one defined in the fragment $\mathcal{X}^{\Uparrow}_{[n,=]}$. In a descendant manner, $RW\_Pred(P)$ is recursively defined over sub-predicates of P as follows:

$-\ \alpha_i = \downarrow$:

$$RW\_Pred(\alpha_i::\eta_i/\cdots/\alpha_k::\eta_k):=$$
$$\downarrow^+::\eta_i[\mathcal{A}^{acc}][RW\_Pred(\alpha_{i+1}::\eta_{i+1}/\cdots/\alpha_k::\eta_k)]/\mathcal{A}^+[1]=\varepsilon::*$$

$-\ \alpha_i \in \{\downarrow^+, \downarrow^*\}$:

$$RW\_Pred(\alpha_i::\eta_i/\cdots/\alpha_k::\eta_k) :=$$
$$\alpha_i::\eta_i[\mathcal{A}^{acc}][RW\_Pred(\alpha_{i+1}::\eta_{i+1}/\cdots/\alpha_k::\eta_k)]$$

$-\ \alpha_i = \varepsilon$:

$$RW\_Pred(\alpha_i::\eta_i/\cdots/\alpha_k::\eta_k) := \varepsilon::\eta_i[RW\_Pred(\alpha_{i+1}::\eta_{i+1}/\cdots/\alpha_k::\eta_k)]$$

As a special case, the predicate $\alpha::\eta/text()='c'$ (text-content comparison) is rewritten, according to the axis α, as follows:

- RW_Pred($\downarrow::\eta/text()=$'c') := $\downarrow^+::\eta[\mathcal{A}^{acc}][self::*/text()=$'c']$/\mathcal{A}^+[1] = \varepsilon::*$
- For $\alpha \in \{\downarrow^+, \downarrow^*\}$, RW_Pred($\alpha::\eta/text()=$'c') := $\alpha::\eta[\mathcal{A}^{acc}]/text()=$'c'
- RW_Pred($\varepsilon::\eta/text()=$'c') := $\varepsilon::\eta/text()=$'c'

Example 12. Consider the access specification of Example 8 and the data view of Fig. 6. It is clear that the predicate $[\underbrace{\downarrow::patient/\downarrow::wardNo = \text{``}421\text{''}}_{P}]$ is sat-

isfied only over the element node $parent_1$. This predicate is rewritten into $[RW\_Pred(P)]$ as follows:

- $[RW\_Pred(P)] = [\downarrow^+::patient[\mathcal{A}^{acc}][RW\_Pred(\downarrow::wardNo=\text{``}421\text{''})]/\mathcal{A}^+[1]=\varepsilon::*]$
- $[RW\_Pred(\downarrow::wardNo=\text{``}421\text{''})] =$
  $[\downarrow^+::wardNo[\mathcal{A}^{acc}][\varepsilon::*/text()=\text{``}421\text{''}]/\mathcal{A}^+[1]=\varepsilon::*]$

Consider the XML document of Fig. 5, it is easy to check that the predicate $[RW\_Pred(P)]$ is satisfied only over the element node $parent_1$.   □

Finally, we generalize the definition of the function *Rewrite* to take into account all queries of the fragment $\mathcal{X}$. Given an access specification $S=(D, ann)$, the function *Rewrite*: $\mathcal{X} \longrightarrow \mathcal{X}^{\Uparrow}_{[n,=]}$ is redefined to rewrite any $\mathcal{X}$ query Q, of the form $\alpha_1::\eta_1[p_1]/\cdots/\alpha_k::\eta_k[p_k]$ $(k \geq 1)$, into another one defined in the fragment $\mathcal{X}^{\Uparrow}_{[n,=]}$ as follows (where $p_i^t=RW\_Pred(p_i)$ for $1 \leq i \leq k$):

$$Rewrite(Q) := \downarrow^*::\eta_k[\mathcal{A}^{acc}][p_k^t][prefix^{-1}(Q)]$$

The qualifier $prefix^{-1}(Q)$ is recursively defined as follows:

- $\alpha_i = \downarrow$:
  $prefix^{-1}(\alpha_1::\eta_1[p_1]/\cdots/\alpha_i::\eta_i[p_i]) :=$
    $\mathcal{A}^{\eta_{i-1}}[p_{i-1}^t][prefix^{-1}(\alpha_1::\eta_1[p_1]/\cdots/\alpha_{i-1}::\eta_{i-1}[p_{i-1}])]$
- $\alpha_i \in \{\downarrow^+, \downarrow^*\}$:
  $prefix^{-1}(\alpha_1::\eta_1[p_1]/\cdots/\alpha_i::\eta_i[p_i]) :=$
    $\alpha_i^{-1}::\eta_{i-1}[p_{i-1}^t][\mathcal{A}^{acc}][prefix^{-1}(\alpha_1::\eta_1[p_1]/\cdots/\alpha_{i-1}::\eta_{i-1}[p_{i-1}])]$
- $\alpha_i = \varepsilon$:
  $prefix^{-1}(\alpha_1::\eta_1[p_1]/\cdots/\alpha_i::\eta_i[p_i]) :=$
    $\varepsilon::\eta_{i-1}[p_{i-1}^t][prefix^{-1}(\alpha_1::\eta_1[p_1]/\cdots/\alpha_{i-1}::\eta_{i-1}[p_{i-1}])]$

As a special case, query of $\mathcal{X}$ of the form $Q_1 \cup \cdots \cup Q_k$ $(k \geq 1)$ is rewritten into *Rewrite*$(Q_1) \cup \cdots \cup$ *Rewrite*(Q_k).

Example 13. Consider the access specification defined in Example 8. The $\mathcal{X}$ query $Q=\downarrow^+::parent/\downarrow::patient[\underbrace{\downarrow::pname = \text{``}Martin\text{''}}_{P}]$ over the data view of Fig. 6 is rewritten over the original data of Fig. 5 as follows:

$$Rewrite(Q)=\downarrow^*::patient[\mathcal{A}^{acc}][RW\_Pred(P)][\uparrow^+::*[\mathcal{A}^{acc}][1]/\varepsilon::parent]$$

$$RW\_Pred(P) = [\downarrow^*::pname[\mathcal{A}^{acc}][\varepsilon::*/text()=\text{``}Martin\text{''}]/\mathcal{A}^+[1]=\varepsilon::*]$$

The evaluation of the query $\mathit{Rewrite}\,(Q)$ over the original data returns the element node $patient_3$ as does the query Q over the data view. $\qquad\square$

We emphasize that the generalization of the function RW_Pred to handle complex predicates is quite straightforward. For instance, $RW\_Pred(P_1 \vee P_2)$ is given by $RW\_Pred(P_1) \vee RW\_Pred(P_2)$. Moreover, $RW\_Pred(P_1[P_2])$ is given by $RW\_Pred(P_1[RW\_Pred(P_2)])$.

5.3 Coping with $\mathcal{X}^{\Uparrow}$ queries

We show how our rewriting function $\mathit{Rewrite}$ can be extended to rewrite the upward axes $\{\uparrow, \uparrow^+, \uparrow^*\}$. Let $S=(D, \mathbf{ann})$ be an access specification. Firstly, the function $\mathit{Rewrite}\colon \mathcal{X}^{\Uparrow} \longrightarrow \mathcal{X}^{\Uparrow}_{[n,=]}$ is redefined to rewrite any $\mathcal{X}^{\Uparrow}$ query Q, of the form $\alpha_1{::}\eta_1[p_1]/\cdots/\alpha_k{::}\eta_k[p_k]$ $(k \geq 1)$, into another one defined in the fragment $\mathcal{X}^{\Uparrow}_{[n,=]}$ as follows (we consider only the case where $\alpha_i \in \{\uparrow, \uparrow^+, \uparrow^*\}$ since the case of the remaining axes is already studied):

$$\mathit{Rewrite}\,(Q) := \downarrow^*{::}\eta_k[\mathcal{A}^{acc}][p_k^t][prefix^{-1}(Q)]$$

The qualifier $prefix^{-1}(Q)$ is recursively defined as follows:

- $\alpha_i = \uparrow$:
  $$prefix^{-1}(\alpha_1{::}\eta_1[p_1]/\cdots/\alpha_i{::}\eta_i[p_i]) :=$$
  $$\downarrow^+{::}\eta_{i-1}[\mathcal{A}^{acc}][p^t_{i-1}][prefix^{-1}(\alpha_1{::}\eta_1[p_1]/\cdots/\alpha_{i-1}{::}\eta_{i-1}[p_{i-1}])]/$$
  $$\mathcal{A}^+[1]{=}\varepsilon{::}\eta_i$

- $\alpha_i \in \{\uparrow^+, \uparrow^*\}$: $(\alpha_i^{-1}{=}\downarrow^+$ if $\alpha_i{=}\uparrow^+$ and $\downarrow^*$ otherwise$)$
  $$prefix^{-1}(\alpha_1{::}\eta_1[p_1]/\cdots/\alpha_i{::}\eta_i[p_i]) :=$$
  $$\alpha_i^{-1}{::}\eta_{i-1}[\mathcal{A}^{acc}][p^t_{i-1}][prefix^{-1}(\alpha_1{::}\eta_1[p_1]/\cdots/\alpha_{i-1}{::}\eta_{i-1}[p_{i-1}])]$$

The function $RW\_Pred\colon \mathcal{X}^{\Uparrow} \longrightarrow \mathcal{X}^{\Uparrow}_{[n,=]}$ is redefined to rewrite any $\mathcal{X}^{\Uparrow}$ predicate P, of the form $\alpha_1{::}\eta_1/\cdots/\alpha_k{::}\eta_k$ $(k \geq 1)$, into another one defined in the fragment $\mathcal{X}^{\Uparrow}_{[n,=]}$ as follows (only the case of upward axes is considered):

- $\alpha_i = \uparrow$:
  $$RW\_Pred(\alpha_i{::}\eta_i/\cdots/\alpha_k{::}\eta_k) := \mathcal{A}^{\eta_i}[RW\_Pred(\alpha_{i+1}{::}\eta_{i+1}/\cdots/\alpha_k{::}\eta_k)]$$
- $\alpha_i \in \{\uparrow^+, \uparrow^*\}$:
  $$RW\_Pred(\alpha_i{::}\eta_i/\cdots/\alpha_k{::}\eta_k) := \alpha_i{::}\eta_i[\mathcal{A}^{acc}][RW\_Pred(\alpha_{i+1}{::}\eta_{i+1}/\cdots/\alpha_k{::}\eta_k)]$$

5.4 Theoretical Results

We present briefly some results that concern the evaluation of the overall answering time of our rewriting approach as well as its correctness.

Lemma 2. *Every $\mathcal{X}^{\Uparrow}_{[n,=]}$ query Q can be evaluated over an XML document T in time $O(|Q|.|T|)$.*

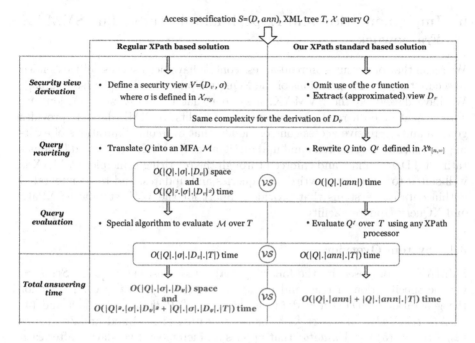

Fig. 7. Comparing our solution with that of [17].

The proof of this lemma is based on the results of the XPath query complexity analysis detailed in [48].

Theorem 2. *Given an access specification $S=(D, \textbf{ann})$, an XML tree $T \in T(D)$ and its virtual view T_v computed w.r.t S. There exists an algorithm* Rewrite *that translates any $\mathcal{X}$ query Q over T_v into an $\mathcal{X}^{\Uparrow}_{[n,=]}$ query Q^t over T at most in time $O(|Q|)$. Moreover, Q^t can be evaluated over T at most in time $O(|Q|.|\textbf{ann}|.|T|)$.*

Theorem 3. *The query rewriting approach is correct for any query of the fragment $\mathcal{X}$.*

Theorem 3 shows the correctness of our query rewriting approach. More specifically, for any access specification $S=(D,\textbf{ann})$, any XML tree $T \in T(D)$ and its virtual view T_v, our rewriting algorithm Rewrite translates any $\mathcal{X}$ query Q over T_v into a safe one Q^t defined over T such that: $\mathcal{S}[\![Q]\!](T_v)=\mathcal{S}[\![Q^t]\!](T)$.

Our algorithm Rewrite and the detailed proofs are given on-line at https://tel.archives-ouvertes.fr/tel-01093661/.

Finally, we make a brief comparison of our XPAth-based solution with that of [17] that is based on Regular XPath. We consider the same access specification, the same XML tree, and we show how an $\mathcal{X}$ query Q over this tree can be answered using both our solution and that of [17]. Figure 7 details the results of this comparison at each step of the XML access control processing.

6   Implementation and Experimental Study: The SVMAX Framework

We recall that our results on read-access control have been successfully extended to secure the update operations of the XQuery Update Facility [31] (see [32,33]). We have developed the SVMAX, a system that facilitates specification and enforcement of both read and update access rights for XML data. It provides general and expressive access control models that overcome limitations of existing approaches. Both of read and update rights of SVMAX are defined by annotating DTD grammars and enforced through the rewriting principle. SVMAX is well-suited to efficiently rewrite such queries and updates, and to be integrated within database systems that provide support for the W3C standards: XPath and XQuery Update Facility.

6.1   System Overview

SVMAX is composed by the following major modules: (1) a *Policy Specifier*, for the specification of read and update privileges; (2) a *View Generator*, for the generation of DTD and data views; (3) an *XPath Rewriter* [49] and (4) an *XQuery Update Rewriter* [33], for the rewriting of read and update queries respectively; (5) the *Validator* that applies an incremental validation after each update operation is performed[12]. These modules are implemented as an API allowing SVMAX to be integrated within existing native XML database systems that are aware of the XML data structure and support W3C standards.

On the other hand, SVMAX can run in standalone mode through its visual tool, SVMAX$^{\mathcal{V}}$. This latter is a GUI tool that monitors the previous modules. More precisely, SVMAX$^{\mathcal{V}}$ is used by the administrator to specify read and update policies, generate virtual views of the DTD and the XML data, and provide these views to the user. The user requests (XPath queries or XQuery update operations) are rewritten, using the adequate rewriter module, to be safely evaluated over the original XML data and then evaluation results are returned to the user. See [35] for more descriptions and screenshots of the system.

We should emphasize that in case of recursive DTDs, the DTD view generation is not always guaranteed [18] or can be of exponential size [50]. More specifically, hiding some information from the DTD may result in a context-free grammar that cannot be captured with a regular grammar[13]. In such situations, our *View generator* module generates an *approximated* DTD view. Our approximations are based on the well-known sufficient conditions for regularization of context-free grammars [51].

6.2   Performance Evaluation

In this section we present an evaluation of SVMAX. Our system is provided both as a Java API and a visual tool, the SVMAX$^{\mathcal{V}}$. Using this latter, one

---

[12] This is still an ongoing work: we deal only with simple kinds of DTDs and update operations, however, the global case is part of our perspective.

[13] It is undecidable in general to find a regular solution for a context-free grammar.

can choose a document DTD, specify access and update policies, and enforce these policies over underlying XML data. We focused in our experiments on the overall-time required for *rewriting* and *evaluation* of XPath queries. The study is conducted on the following aspects: (1) measure of scalability and degradation of our rewriting approaches, and (2) comparison of SVMAX with respect to naive approach in terms of overall answering time. Since our system can be integrated within existing NXDs, the other concern of experimentation is (3) a study of the integration efficiency.

(1) Scalability. We measure the time required by SVMAX to rewrite general XPath queries. We use the complex real-life recursive DTD **GedML**[14] and we generate randomly 10 access specifications by varying the number of annotations (from 20 into 200). After, we define different XPath queries of size[15] 400 that include most features of the XPath fragment $\mathcal{X}^{\Uparrow}$: with $\downarrow^*$-axis (Q_1); with $\downarrow^*$-axis and predicates (Q_2); with $\downarrow$-axis (Q_3); with $\downarrow$-axis and predicates (Q_4); with $\downarrow^*$-axis, predicates, and $*$-labels inside predicates (Q_5); with $\downarrow$-axis, predicates, and $*$-labels inside predicates (Q_6). Note that the used predicates contain different operators (e.g. $\wedge$, $\vee$, and text comparison).

Using SVMAX, we rewrite these queries according to each of the access specifications previously generated. Figure 8 shows the overall rewriting times. Notice that the rewriting time obtains a constant nature, i.e., it does not increase with the growth of the number of access annotations. This can be explained by the fact that, for an XPath query in input, our rewriter parses all its sub-queries (with the form *axis::label*) and rewrites them using the *accessibility predicate*. The computation time of this latter is negligible (less than 10 ms for large access specifications), and thus, our rewriting time depends basically on the parsing of the query, then on the size of the query. Since our queries have the same size, the overall rewriting time does not depend on the number of access annotations and still remains constant at some point. Moreover, we remark that in general, a query with $\downarrow^+$-axis requires more rewriting time than a query with $\downarrow$-axis (Q_1 w.r.t Q_3), also a query with predicates consumes some additional time (Q_2 w.r.t Q_1; and Q_4 w.r.t Q_3). The $*$-labels require less rewriting time (Q_2 w.r.t Q_5; and Q_4 w.r.t Q_6).

(2) Policy Enforcement. We measure the end-to-end processing time of our system for larger access specifications and general XPath queries. Since no tool exists in practice to secure querying of recursive XML views, we compare our system only w.r.t some naive approach as explained in the following.

We generate an XML document T of size 10MB that conforms to the GedML DTD, and different access specification $S^i=(GedML, ann)$ of size i ($i=|ann|$), where i is varying from 10 to 150. We define after a complex XPath query with

---

[14] Genealogy Markup Language: http://xml.coverpages.org/gedml-dtd9808.txt.
[15] The size of an XPath expression is the occurrence number of all its element types, $*$-labels, and text() functions.

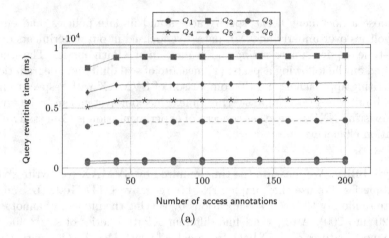

Fig. 8. SVMAX rewriting degradation for read update rights.

important size, different axes and complex predicates. This query is rewritten, w.r.t each specification S^i, both with our approach and using the on-the-fly materialization [24] as the naive approach. Figure 9 shows the answering times of each approach[16]. It is clearly shown that in case of large size of specifications and XML data, our system requires a small answering time and achieves an improvement of the naive approach by up to a factor of 10.

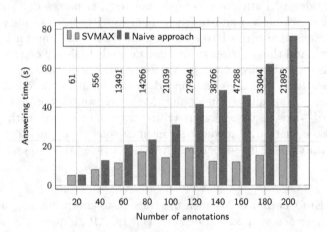

Fig. 9. Overall answering time: SVMAX versus naive approach.

(3) Integrating SVMAX Within NXDs. Finally, we use SVMAX as a simple Java API and we integrate it within different native XML databases: (1) *BaseX*, (2) *Sedna* and (3) *eXist*. The selection of these NXDs is done according to their growing use, as well as to their supports for querying and updating

---

[16] In the following figures, the numbers of queried nodes are depicted at the middle.

of XML data. The XPath language is supported by the three NXDs. However, only *BaseX* provides implementation for the XQuery update facility; each of the remaining systems provides a proprietary update language.

The communication between the SVMAX API and the underlying database system is ensured by using the APIs XQJ and XML:DB, present in most systems. The goal of this integration is to offer existing databases easy-to-use and efficient support to securely manipulate (recursive) XML views, as well as to leverage advantages of these systems (e.g. query optimization technologies).

We generate a simple XML document of 2 MB, a general query, and some policies P^1,...,P^{10} defined with the same principle explained in the previous subsection. Using the SVMAX rewriters, the query is safely rewritten w.r.t the different policies and sent to the underlying database for evaluation. The overall answering times (rewriting and evaluation) are depicted in Fig. 10. We remark that *eXist* database takes more time than the other (282 s for the simple policy P^1, i.e., with 20 annotations). The *BaseX* XQuery processor overcomes noticeably the *Sedna* processor in general by up to a factor of 2.

The first result of this study shows that our system has been successfully and easily integrated within such database systems. Since there are various implementation of the W3C standards, the other benefit of this study is to know with which XPath (resp. XQuery) processor the SVMAX rewritten queries may provide more efficiency.

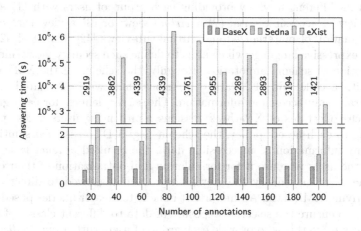

Fig. 10. Integration of SVMAX within NXDs.

7   Related Work

Figure 11 summaries the evolution of the XML access control models during the two decades. At the outset, used approaches [47,52] consisted on annotating naively the XML data with some security labels to specify which actions can be performed on which XML nodes, and thus restrict access to sensitive data through these labels. Although, some improvements [41,53] have been made in order to avoid the costly re-annotation of the data, these naive approaches are time consuming and generally difficult to apply for example in case of different users, multiple actions, and dynamic policies. Other models have been proposed [34,46] that define access policies without any labeling of data, and enforce these policies during the evaluation of users requests (read-access queries or update operations). An access policy is defined as a set of XPath expressions, each one refers to a set of XML nodes over which the user can execute some actions (read or update). The users requests are *rewritten* w.r.t the underlying access policies by adding some XPath predicates in order to execute the requested action only on authorized data (i.e. data that can be queried and/or updated). These XPath-based approaches outperform the instance-based approaches in most cases. However, the major limitation of these models is the lack of support for authorized users to access the data: the schema of accessible data is necessary for the users in order to formulate and optimize their queries; as well as for the security administrator for understanding how the authorized view of the XML data, for a group of users, will actually look like.

To overcome limitations of node-labeling protection and XPath-based protection, Stoica and Farkas [40] introduced the notion of *XML security view* that consists on defining, for each group of users, a view of the XML document that displays all and only accessible information. This notion has been refined later and used in different ways by providing each group of users with (1) a *materialized* view of accessible data; (2) a *virtual* view; or (3) a view that consists of a combination of materialized and virtual sub-views [42]. Fan et al. [25] proposed an expressive language which aims to define such security views and based on the notion of schema annotation. Roughly, the schema of the XML data is paired with a collection of XPath expressions that, when evaluated over the data, extract only accessible information. The server defines, for each group of users, such collections of XPath expressions representing users access policies. According to each access policy, the schema (e.g. a DTD) is then sanitized by eliminating information of inaccessible data, the resulted *schema view* is provided to the users who use it for formulation and optimization of their queries. While the users may query the views, they are not allowed to directly query the underlying XML data. An important issue is to answer queries posed on the views and to ensure the selective exposure of data to different classes of users.

One way to do this is to provide each group of users with a *materialized* view of all and only accessible data (as studied in [24]), which is used to evaluate users queries directly over it and offers faster access to the data. However, when the XML data and/or the access policies are changed, all users views should be (incrementally) maintained [55–58]. Note that in some cases, incremental maintenance of materialized views leads to the same performances as re-computation of the views from scratch. In addition to the maintenance cost, materialization of all users views within the server is time and memory consuming.

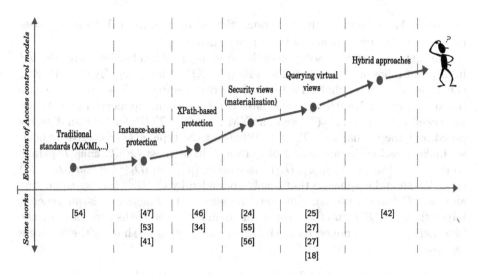

Fig. 11. Evolution of XML Access Control Models.

The *view virtualization* is the adequate and more scalable solution in case of huge data, an important number of users, and dynamic policies. Fan et al. [25] defined the notion of *query rewriting* that consists on translating queries posed over virtual views into equivalent ones to be evaluated over the original data. Since DTDs found in practice are often recursive [26], many authors have refined this work to use more expressive query language [17,18,27], namely Regular XPath. Regular XPath is more expressive than XPath and allows definition of recursive paths. The use of this language to secure XML data has been more studied in [17,18]. However, Regular XPath based solutions still a theoretical achievement and may be impractical since rewriting of Regular XPath expressions may be of exponential cost as we have shown in Sect. 5.4. In addition, Regular XPath is not commonly used in practice[17] and most of the commercial database systems (e.g. eXistdb) offer support for the W3C standards: XPath and XQuery. Thus, the securing of such queries remains a strong necessity.

8   Conclusions

We aimed to provide a practical solution for the open problem that consists on rewriting XPath queries under DTDs recursion. We have investigated the extension of the downward class of XPath with some axes and operators, and showed that the resulted XPath fragment $\mathcal{X}_{[n,=]}^{\Uparrow}$ can be used to rewrite efficiently any $\mathcal{X}$ query, over the data view, into a safe one that can be evaluated directly over the original data. Our proposal yields the first practical solution for the rewriting problem. The conducted experimentation shows the efficiency of our

---

[17] Note that no tool exists in practice to evaluate Regular XPath queries.

approach. Most importantly, the translation of queries from $\mathcal{X}$ to $\mathcal{X}_{[n,=]}^{\Uparrow}$ does not impact the performance of the queries answering.

Recall that a previous solution of the rewriting problem has been investigated in [27] that relies on the non-standard Regular XPath language. By the following comparison, we show that XPath-based rewriting is more efficient than the one based on Regular XPath since this later can lead to an exponential cost. Given an access specification $S=(D, \boldsymbol{ann})$, an XML tree $T \in \mathcal{T}(D)$, let Q be an $\mathcal{X}$ query posed over the virtual view T_v of T. Whatever the type of D (recursive or not), we make possible the answering of Q over T in at most $O(|Q|.|\boldsymbol{ann}|.|T|)$ time, while [17] do this in $O(|Q|.|\sigma|.|D_v|)$ space and $O(|Q|^2.|\sigma|.|D_v|^2+|Q|.|\sigma|.|D_v|.|T|)$ time. We should emphasize that $|\boldsymbol{ann}|$ is bounded by $O(|D|^2)$ (i.e. we can define at most $|D|^2$ annotations). However, the size of the function σ is, in general, larger than $O(|D|^2)$. In other words, the number of the paths presented by the function σ may be exponential on the size of the DTD as we show by the following example.

Example 14. Consider the DTD $D=(\{Root, A_1, \ldots, A_n\}, P, Root)$ where $n \in \mathbb{N}$ and the production rules are given as follows:

$$P(Root) := (A_1| \cdots |A_n)$$
$$P(A_i) := (A_1| \cdots |A_{i-1}|A_{i+1}| \cdots |A_n), i \leq n$$

We define now the access specification $S=(D, \boldsymbol{ann})$ where $\boldsymbol{ann}$ contains only the default annotation $ann(Root)=Y$, i.e. all element types of D are accessible. It is easy to prove that, for any element types A_i, A_j ($i \leq n$ and $j \leq n$), the number of paths presented by $\sigma(A_i, A_j)$ may be bounded by: $\Sigma_{1 \leq i \leq n-2} \frac{(n-2)!}{(n-2-i)!}$.   □

Finally, we conclude that our rewriting approach is more efficient in practice than the one based on Regular XPath and requires an answering time that is linear on the size of the input query, the number of annotations, and the size of the XML data. This would lead for an efficient integration of our solution within some existing database systems. Moreover, by working with the XPath standard, we make possible the use of a bulk of interesting results found around the XPath language (e.g. XPath queries optimization [59,60] and efficient evaluation [61]).

References

1. Robie, J., Chamberlin, D., Dyck, M., Florescu, D., Melton, J., Siméon, J.: Extensible Markup Language (XML) 1.0 (Fifth Edition). W3C Recommendation (2008). http://www.w3.org/TR/2008/REC-xml-20081126/
2. Bray, T., Paoli, J., Sperberg-McQueen, C.M., Maler, E., Yergeau, F., Cowan, J.: Extensible Markup Language (XML) 1.1 (Second Edition). W3C Recommendation (2006). http://www.w3.org/TR/2006/REC-xml11-20060816/
3. Amavi, J., Chabin, J., Halfeld-Ferrari, M., Réty, P.: A toolbox for conservative XML schema evolution and document adaptation. In: Decker, H., Lhotská, L., Link, S., Spies, M., Wagner, R.R. (eds.) DEXA 2014, Part I. LNCS, vol. 8644, pp. 299–307. Springer, Heidelberg (2014)

4. Chabin, J., Halfeld Ferrari, M., Musicante, M.A., Réty, P.: Conservative type extensions for XML data. In: Hameurlain, A., Küng, J., Wagner, R. (eds.) TLDKS IX. LNCS, vol. 7980, pp. 65–94. Springer, Heidelberg (2013)
5. Gerald, B., Sleeper, H., Gregorowicz, A., Dingwell, R.: hData - a simple XML framework for health data exchange. In: Proceedings of Balisage: The Markup Conference, Montral, Canada, August 11–14, 2009, vol. 3, pp. 299–307 (2009)
6. Fried, E., Geng, Y., Ullrich, S., Kneer, D., Grottke, O., Rossaint, R., Deserno, T.M., Kuhlen, T.: MEDOX: an XML-based approach of medical data organization for segmentation and simulation. In: Bildverarbeitung für die Medizin 2010 - Algorithmen - Systeme - Anwendungen, Aachen, Germany, March 14–16, 2010. CEUR Workshop Proceedings, vol. 574, 251–255. CEUR-WS.org (2010)
7. Cavalini, L.T., Cook, T.W.: Use of XML schema definition for the development of semantically interoperable healthcare applications. In: Gibbons, J., MacCaull, W. (eds.) FHIES 2013. LNCS, vol. 8315, pp. 125–145. Springer, Heidelberg (2014)
8. la Rosa Algarin, A.D., Demurjian, S.A., Berhe, S., Pavlich-Mariscal, J.A.: A security framework for XML schemas and documents for healthcare. In: 2012 IEEE International Conference on Bioinformatics and Biomedicine Workshops, BIBMW 2012, Philadelphia, USA, October 4–7, 2012, pp. 782–789. IEEE (2012)
9. Steele, R., Gardner, W., Chandra, D., Dillon, T.S.: Framework and prototype for a secure XML-based electronic health records system. IJEH 3(2), 151–174 (2007)
10. Kumar, C.S., Govardhan, A., Rao, C.V.G.: Usage of XML technology in electronic health record for effective heterogeneous systems integration in healthcare. IJMEI 1(4), 399–406 (2009)
11. Thuy, P.T.T., Lee, Y., Lee, S.: Semantic and structural similarities between XML schemas for integration of ubiquitous healthcare data. Pers. Ubiquit. Comput. 17(7), 1331–1339 (2013)
12. IBM jStart team: IBM Emerging Technology's client engagement team. http://www-01.ibm.com/software/ebusiness/jstart/
13. DITA OASIS Standard: An XML architecture for designing, writing, managing, and publishing information. http://dita.xml.org/
14. ebXML consortium: Electronic Business using eXtensible Markup Language. http://www.ebxml.org/
15. Oracle White Paper: Sun Storage 7000 Unified Storage Systems and XML-Based Archiving for SAP Systems, April 2010. http://www.oracle.com/us/solutions/sap/database/ss7000-sap-implementation-guide-352637.pdf
16. Rassadko, N.: Policy classes and query rewriting algorithm for XML security views. In: Damiani, E., Liu, P. (eds.) Data and Applications Security 2006. LNCS, vol. 4127, pp. 104–118. Springer, Heidelberg (2006)
17. Fan, W., Geerts, F., Jia, X., Kementsietsidis, A.: Rewriting regular xpath queries on XML views. In: ICDE, pp. 666–675. IEEE (2007)
18. Groz, B., Staworko, S., Caron, A.-C., Roos, Y., Tison, S.: XML security views revisited. In: Gardner, P., Geerts, F. (eds.) DBPL 2009. LNCS, vol. 5708, pp. 52–67. Springer, Heidelberg (2009)
19. Luo, B., Lee, D., Lee, W.C., Liu, P.: Qfilter: rewriting insecure XML queries to secure ones using non-deterministic finite automata. VLDB J. 20(3), 397–415 (2011)
20. Cong, G.: Query and update through XML views. In: Bhalla, S. (ed.) DNIS 2007. LNCS, vol. 4777, pp. 81–95. Springer, Heidelberg (2007)
21. Damiani, E., Fansi, M., Gabillon, A., Marrara, S.: A general approach to securely querying XML. Comput. Stand. Interfaces 30(6), 379–389 (2008)

22. Clark, J., DeRose, S.: XML path language (XPath) 1.0. W3C Recommendation, November 1999. http://www.w3.org/TR/xpath/
23. Berglund, A., Boag, S., Chamberlin, D., Fernández, M.F., Kay, M., Robie, J., Siméon, J.: XML path language (XPath) 2.0 (second edition). W3C Recommendation, December 2010. http://www.w3.org/TR/2010/REC-xpath20-20101214/
24. Kuper, G.M., Massacci, F., Rassadko, N.: Generalized XML security views. Int. J. Inf. Sec. **8**(3), 173–203 (2009)
25. Fan, W., Chan, C.Y., Garofalakis, M.N.: Secure XML querying with security views. In: Proceedings of the ACM SIGMOD International Conference on Management of Data, pp. 587–598. ACM (2004)
26. Choi, B.: What are real dtds like? In: Fifth International Workshop on the Web and Databases (WebDB), pp. 43–48 (2002)
27. Fan, W., Geerts, F., Jia, X., Kementsietsidis, A.: SMOQE: a system for providing secure access to XML. In: Proceedings of the 32nd International Conference on Very Large Data Bases, pp. 1227–1230. ACM (2006)
28. Marx, M.: XPath with conditional axis relations. In: Bertino, E., Christodoulakis, S., Plexousakis, D., Christophides, V., Koubarakis, M., Böhm, K. (eds.) EDBT 2004. LNCS, vol. 2992, pp. 477–494. Springer, Heidelberg (2004)
29. Wood, P.T.: Containment for XPath fragments under DTD constraints. In: Calvanese, D., Lenzerini, M., Motwani, R. (eds.) ICDT 2003. LNCS, vol. 2572, pp. 297–311. Springer, Heidelberg (2002)
30. Neven, F., Schwentick, T.: On the complexity of Xpath containment in the presence of disjunction, DTDs, and variables. Logical Methods in Computer Science 2(3) (2006)
31. Robie, J., Chamberlin, D., Dyck, M., Florescu, D., Melton, J., Siméon, J.: Xquery update facility 1.0. W3C Recommendation, March 2011. http://www.w3.org/TR/xquery-update-10/
32. Mahfoud, H., Imine, A.: A general approach for securely updating XML data. In: Proceedings of the 15th International Workshop on the Web and Databases (WebDB 2012), pp. 55–60 (2012)
33. Mahfoud, H., Imine, A.: On securely manipulating XML data. In: Garcia-Alfaro, J., Cuppens, F., Cuppens-Boulahia, N., Miri, A., Tawbi, N. (eds.) FPS 2012. LNCS, vol. 7743, pp. 293–307. Springer, Heidelberg (2013)
34. Fundulaki, I., Maneth, S.: Formalizing XML access control for update operations. In: SACMAT, pp. 169–174. ACM (2007)
35. Mahfoud, H., Imine, A., Rusinowitch, M.: SVMAX: a system for secure and valid manipulation of XML data. In: Proceedings of the 17th International Database Engineering & Applications Symposium (IDEAS), pp. 154–161. ACM (2013)
36. Jia, X.: From Relations to XML: Cleaning, Integrating and Securing Data. Doctor of philosophy, Laboratory for Foundations of Computer Science. School of Informatics. University of Edinburgh (2007)
37. Fan, W., Yu, J.X., Li, J., Ding, B., Qin, L.: Query translation from XPath to SQL in the presence of recursive dtds. VLDB J. **18**(4), 857–883 (2009)
38. Krishnamurthy, R., Chakaravarthy, V.T., Kaushik, R., Naughton, J.F.: Recursive XML schemas, recursive XML queries, and relational storage: XML-to-SQL query translation. In: Proceedings of the 20th International Conference on Data Engineering (ICDE 2004), pp. 42–53. IEEE Computer Society (2004)
39. ten Cate, B.: The expressivity of XPath with transitive closure. In: Proceedings of the Twenty-Fifth ACM SIGACT-SIGMOD-SIGART Symposium on Principles of Database Systems (PODS 2006), pp. 328–337. ACM (2006)

40. Stoica, A., Farkas, C.: Secure XML views. In: Research Directions in Data and Applications Security, IFIP WG 11.3 Sixteenth International Conference on Data and Applications Security. IFIP Conference Proceedings, vol. 256, pp. 133–146. Kluwer (2002)

41. Duong, M., Zhang, Y.: An integrated access control for securely querying and updating XML data. In: Proceedings of the Nineteenth Australasian Database Conference (ADC). CRPIT, vol. 75, pp. 75–83. Australian Computer Society (2008)

42. Thimma, M., Tsui, T.K., Luo, B.: HyXAC: a hybrid approach for XML access control. In: 18th ACM Symposium on Access Control Models and Technologies (SACMAT), ACM (2013)

43. Fegaras, L.: Incremental maintenance of materialized XML views. In: Hameurlain, A., Liddle, S.W., Schewe, K.-D., Zhou, X. (eds.) DEXA 2011, Part II. LNCS, vol. 6861, pp. 17–32. Springer, Heidelberg (2011)

44. Shastry, P.D.N.M.: Integrated Healthcare IHE Pathway for the Patients: Patient Treatment Lifecycle Management (PTLM). Radiology Clinic, United Kingdom (2000). (October 2012) http://www.clinrad.nhs.uk/

45. Samarati, P., di Vimercati, S.C.: Access control: policies, models, and mechanisms. In: Focardi, R., Gorrieri, R. (eds.) FOSAD 2000. LNCS, vol. 2171, pp. 137–146. Springer, Heidelberg (2001)

46. Fundulaki, I., Marx, M.: Specifying access control policies for XML documents with XPath. In: SACMAT 2004, 9th ACM Symposium on Access Control Models and Technologies, pp. 61–69, ACM (2004)

47. Murata, M., Tozawa, A., Kudo, M., Hada, S.: XML access control using static analysis. ACM Trans. Inf. Syst. Secur. 9(3), 292–324 (2006)

48. Gottlob, G., Koch, C., Pichler, R.: Efficient algorithms for processing XPath queries. ACM Trans. Database Syst. 30(2), 444–491 (2005)

49. Mahfoud, H., Imine, A.: Secure querying of recursive XML views: a standard XPath-based technique. In: WWW (Companion Volume), pp. 575–576. ACM (2012)

50. Kuper, G.M., Massacci, F., Rassadko, N.: Generalized XML security views. In: 10th ACM Symposium on Access Control Models and Technologies (SACMAT), pp. 77–84. ACM (2005)

51. Andrei, S., Chin, W.N., Cavadini, S.V.: Self-embedded context-free grammars with regular counterparts. Acta Inf. 40(5), 349–365 (2004)

52. Murata, M., Tozawa, A., Kudo, M., Hada, S.: XML access control using static analysis. In: Proceedings of the 10th ACM Conference on Computer and Communications Security (CCS), pp. 73–84. ACM (2003)

53. Duong, M., Zhang, Y.: Dynamic labelling scheme for XML data processing. In: Meersman, R., Tari, Z. (eds.) OTM 2008, Part II. LNCS, vol. 5332, pp. 1183–1199. Springer, Heidelberg (2008)

54. Oasis extensible access control markup language (XACML) TC, January 3013. https://www.oasis-open.org/committees/tc_home.php?wg_abbrev=xacml

55. Bonifati, A., Goodfellow, M.H., Manolescu, I., Sileo, D.: Algebraic incremental maintenance of XML views. In: 14th International Conference on Extending Database Technology (EDBT), pp. 177–188. ACM (2011)

56. Nica, A.: Incremental maintenance of materialized views with outerjoins. Inf. Syst. 37(5), 430–442 (2012)

57. Gupta, A., Mumick, I.S.: Maintenance of materialized views: Problems, techniques, and applications. IEEE Data Eng. Bull. 18(2), 3–18 (1995)

58. Gupta, A., Mumick, I.S., Rao, J., Ross, K.A.: Adapting materialized views after redefinitions: techniques and a performance study. Inf. Syst. 26(5), 323–362 (2001)

59. Maneth, S., Nguyen, K.: XPath whole query optimization. PVLDB **3**(1), 882–893 (2010)
60. Georgiadis, H., Charalambides, M., Vassalos, V.: A query optimization assistant for XPath. In: Proceedings of the 14th International Conference on Extending Database Technology (EDBT 2011), ACM (2011)
61. Hsu, W.C., Liao, I.E.: CIS-X: a compacted indexing scheme for efficient query evaluation of XML documents. Inf. Sci. **241**, 195–211 (2013)

Increasing Coverage in Distributed Search and Recommendation with Profile Diversity

Maximilien Servajean[1]([✉]), Esther Pacitti[1], Miguel Liroz-Gistau[2],
Sihem Amer-Yahia[3], and Amr El Abbadi[4]

[1] INRIA & LIRMM, University of Montpellier, Montpellier, France
{servajean,pacitti}@lirmm.fr
[2] INRIA & LIRMM, Montpellier, France
miguel.liroz_gistau@inria.fr
[3] CNRS, LIG, Grenoble, France
sihem.amer-yahia@imag.fr
[4] Department of Computer Science, University of California, Santa Barbara, USA
amr@cs.ucsb.edu

Abstract. With the advent of Web 2.0 users are producing bigger and bigger amounts of diverse data, which are stored in a large variety of systems. Since the users' data spaces are scattered among those independent systems, data sharing becomes a challenging problem. Distributed search and recommendation provides a general solution for data sharing and among its various alternatives, gossip-based approaches are particularly interesting as they provide scalability, dynamicity, autonomy and decentralized control. Generally, in these approaches each participant maintains a cluster of "relevant" users, which are later employed in query processing. However, as we show in the paper, only considering relevance in the construction of the cluster introduces a significant amount of redundancy among users, which in turn leads to reduced recall. Indeed, when a query is submitted, due to the high similarity among the users in a cluster, the probability of retrieving the same set of relevant items increases, thus limiting the amount of distinct results that can be obtained.In this paper, we propose a gossip-based search and recommendation approach that is based on diversity-based clustering scores. We present the resultant new gossip-based clustering algorithms and validate them through experimental evaluation over four real datasets, based on *MovieLens-small*, *MovieLens*, *LastFM* and *Delicious*. Compared with state of the art solutions, we show that taking into account diversity-based clustering score enables to obtain major gains in terms of recall while reducing the number of users involved during query processing.

1 Introduction

In the context of Web 2.0, users become massive producers of diverse data (*e.g.* photos, videos, scientific data) that can be stored in a large variety of systems

---

Work conducted within the Institut de Biologie Computationnelle and partially funded by the labex NUMEV and the CNRS project Mastodons.

A. Hameurlain et al. (Eds.): TLDKS XXII, LNCS 9430, pp. 115–144, 2015.
DOI: 10.1007/978-3-662-48567-5_4

(*e.g.* Dropbox, Facebook, Flickr, Google+, local computer or smartphone). Users are often willing to share their data with other users in a community of interest. However, the fact that their data spaces are distributed in many different systems makes data sharing especially difficult. For instance, an artist photographer who wants to share her pictures within an online community of photographers may have to log in several different Web applications such as deviantArt, Facebook or Flickr, each with a different interface and account. Similarly, a scientist who needs to search for scientific datasets within an online community of scientists will be faced with the problem that the relevant data is typically distributed in many different labs' servers or scientists' local computers. Furthermore, since this data is hidden to web crawlers, traditional search engines become useless. In order to mitigate this problem, some Web applications allow grouping several accounts and data from different systems (*e.g.* Facebook enables to regroup Dropbox and blogs into a single Facebook account). However, they are limited to a few well-known systems.

In this context of large scale distribution of users and data, a general solution to data sharing is offered by distributed search and recommendation [1,2]. In this paper, we adopt a peer-to-peer gossip-based approach, because it provides important properties such as scalability, dynamicity, autonomy and decentralized control [3]. Within an online community, each user u is associated to a virtual data space that contains all the data items (stored in different systems) it shares. Given u and a keyword query q, the goal of our search and recommendation approach is to recommend to u items that are relevant with respect to q and that are shared by other users, regardless of the systems that store the items. Then, a recommended item is simply a reference that can be used to retrieve the actual data item. In other words, we combine search and recommendation in the sense that a user u searches relevant items among those recommended by users similar to u.

Distributed search and recommendation has received considerable attention [1,2,4,5]. However, one open problem is the ability to attain high recall results. A query is generally forwarded only to a subset of users who will be employed to process queries and return recommendations. To compute this subset of users, many solutions cluster relevant user profiles implicitly using gossip protocols. Gossip protocols are known to be highly resilient, scalable and converge quickly [3], which makes them a good alternative for distributed search and recommendation. A *User Network* (*U-Net* in the following) refers to the cluster of relevant users, a user u is aware of by gossiping, using a score (*e.g.* similarity between u and the users in *U-Net*). At each gossip round, the most relevant users are kept in *U-Net*. Since *U-Net* is used to guide recommendations given a keyword query, the relevance score used in the clustering process plays a very important role to increase the number of relevant items retrieved with respect to the whole set of items (*i.e.* recall), known as the global corpus.

Relevance scores (*e.g.* Jaccard, overlap) define how well a user profile v meets the needs of another user u. Most of the existing solutions exploit different kinds of relevance scores to increase recall [2,4–7]. But recall results remain limited.

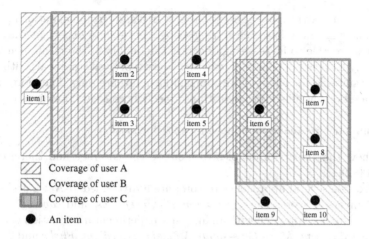

Fig. 1. Example of the coverage of three users.

Two main reasons can be highlighted to explain why recall results remain limited:

1. **Profile redundancy:** using relevance as the clustering score introduces a significant amount of redundant user profiles in each user's *U-Net*. As a result, when a query is submitted, since many user profiles in *U-Net* are quite similar (*i.e.* redundant), and these users are chosen to provide recommendations to answer the query, the probability of retrieving the same set of relevant items increases and recall results remain low.
2. **Network redundancy:** using relevance as the clustering score introduces a significant amount of redundancy between the *U-Net* of similar users. As a result, from the second hop on, queries have a high probability to be forwarded back to users that have already been queried, and recall results remain low.

In *Information Retrieval,* **diversity** is often combined to **relevance** in a score called **usefulness** to overcome redundancy between the items of a result list [8,9]. In our context, we claim that diversity should be used when clustering user profiles in *U-Net*, instead of just relevance. This way, a more diverse set of results will be returned from different users and the recall would be enhanced.

Example 1. Figure 1 shows an example of the benefit of combining **relevance** with **diversity**. Suppose that the relevance metric is the number of items and that the *U-Net* size if 2. Each rectangle represents the items shared by a given user. We want to chose 2 of the users so that the number of different accessible items is maximized. Considering only relevance (in the example, the number of items known by a given user) would result in a *U-Net* composed of users A and C enabling to retrieve 8 out of 10 items. However, any other combinations (*i.e.* A and B or B and C) would have enabled to retrieve at least 9 out of 10 items. This is due to the redundancy of A with respect to C (5 of the items are shared by both users and only 1 is shared only by A and 2 by C).

In this paper, we propose a gossip-based search and recommendation approach based on new diversity-based clustering scores. As we show experimentally, these new scores are able to increase significantly the quality of the recommendations returned by the system. However, existing peer-to-peer clustering algorithms are no longer suitable since they are optimized for relevance only. Therefore we also propose a set of new clustering algorithms especially conceived for diversity-based scores.

In summary, we make the following contributions:

1. We show that *diversity* enables to increase the coverage and therefore, the recall [8,9].
2. We propose several approaches to compute a diversified *U-Net* based on different diversity scores, namely *randomness*, *usefulness* and *coverage*.
3. We validate our approach with an intensive experimental evaluation using four different datasets: *MovieLens-small*, *MovieLens-medium*, *Flickr* and *LastFM*. We observe that diversification enables a huge increase of recall regardless of the relevance score used. Compared with state of the art solutions, we obtain an excellent gain with recall results up to 70% times higher compared to similarity-based approaches and up to 22 times better than random-based approaches while involving a very small number of users.

This paper is a major extension of [10], with more than 50% of new material, including new diversification techniques for computing *U-Net* supported by new experiments. More precisely, in Sect. 4 we present three categories of algorithms to compute a diversified *U-Net*: *random-based*, *usefulness-based* and *coverage-based*. Then, in Sect. 5, we report intensive experiments showing the benefits of our techniques.

This paper is organized as follows. Section 2 provides some basic concepts and gives the problem definition. In Sect. 3, we describe our new clustering score based on the notion of *neighbor's coverage* and *usefulness*. In Sect. 4 we present the details of our algorithms that maintain a diversified *U-Net*. In Sect. 5, we provide an experimental evaluation. In Sect. 6, we compare our contributions with related work. Finally, Sect. 7 concludes and provides directions for future work.

2   Basic Concepts and Problem Definition

In this section, we introduce the background necessary to understand the problem we address.

In our distributed search and recommendation approach, whenever a user u submits a query q, the system sends q to a subset of users that we call *U-Net*, who will return their relevant results to u and will also recursively forward the query to the users in their *U-Net* until the *TTL* (*i.e.* Time-to-Live) is reached. *TTL* defines the maximum number of hops (*i.e.* recursion) the query can do on the network. To build *U-Net*, we use a two-step approach. First, based on *random gossiping* each user u is aware of other peers available on the network.

Second, by means of a *clustering* algorithm, u chooses among these users the best ones to answer u's queries and keep them in *U-Net*.

More precisely, our peer-to-peer model is expressed based on a graph $G = (U, I, E)$, where $U = \{u_1, ..., u_n\}$ is the set of users distributed over the network, $I = \{i_1, ..., i_m\}$ the set of shared data items (in the following, an item refers to a data item), and $E = \{e_1, ..., e_k\}$ the set of directed edges among users and between users and items. This model is very generic. In our case, users are independent nodes in the network. A node can be a physical computer or a virtual node in a server.

Definition 1 (*U-Net*). *Given a user u, its User Network, or U-Net, refers to the cluster of relevant users u is aware of. There is an edge $e(u,v)$ in the graph between u and a user v, if and only if v is in u's U-Net.*

With random gossiping [3], each user keeps locally a random view of its dynamic acquaintances (or view entries). Each view entry corresponds to a user profile. Periodically, each user chooses randomly a contact (view entry) to gossip with. The two involved users then exchange a subset of each other's view (*i.e.* user profiles), and update their view state. Then, in state-of-the-art solutions, after each gossip exchange, the random view is used to update the *U-Net* if more relevant (*e.g.* Jaccard) profiles are found in the updated view. Table 1 presents possible metrics to evaluate the relevance of a user v with respect to a user u:

Table 1. Relevance metrics used to build *U-Net*

overlap(u, v) $= |I_u \cap I_v|$	**over_big(u, v)** $= |I_u \cap I_v| + |I_v|$
Jaccard(u, v) $= \frac{|I_u \cap I_v|}{|I_u \cup I_v|}$	**cosine(u, v)** $= \frac{\overrightarrow{I_u} \cdot \overrightarrow{I_v}}{\|\overrightarrow{I_u}\| \, \|\overrightarrow{I_v}\|}$

where I_u and I_v are the items shared by u and v, respectively, and $\overrightarrow{I}_u$ and $\overrightarrow{I}_v$ are a vectorial representation of u and v's profiles respectively. For instance $overlap(u, v)$ will attribute high relevance scores to users that share the same items and $Jaccard(u, v)$ will take into account the number of items both users share (*i.e.* intersection) with respect to the total number (*i.e.* union) of items they share. Notice that other similarity metrics (*e.g.* Dice [11], Tanimoto [12]) that are not tested in this paper result in the same limitation: clusters of similar users are very redundant since they are composed of users that are very similar to the current one.

Items are represented based on their meta-data's attributes (*e.g.* the keywords contained in the item's title or its tags); these attributes can be evaluated through a vector space model [13].

Each *user profile* is defined based on the items the user shares, I_u. As presented previously, several relevance score (*e.g. Jaccard, cosine*) can be used with good results.

As mentioned before, whenever a user u submits a keywords query $q = k_1, ..., k_w$, the query is redirected to all users in the participating users' *U-Net* recursively, until a predefine upper threshold, TTL. Whenever a user v receives a query, it computes its *top-k* most relevant items with respect to the query using a specific relevance score – not necessarily the same used in *U-Net* construction. Then, v returns them to u. A recommended item is defined by its identifier, its content (*e.g.* title), v's identifier and v's profile. Once u receives the set of recommended items from $v_1, ..., v_n$ with respect to its query q, it ranks them based on their relevance with respect to the query:

$$Rec_q = rank(rec_q^1(it_1, ...) \cup ... \cup rec_q^n(it_p, ...)) \tag{1}$$

where $rec_q^1(it_1, ...)$ is a recommendation (*i.e.* a set of recommended items) coming from a user v_1. The relevance between an item and a query can computed using similar metrics than those presented in Table 1. For instance, *Jaccard* can be used to evaluated the number of common attributes (*i.e.* query's keywords and item's attribute) between the query and the item with respect to their total number of attributes.

To evaluate the quality of search and recommendation, we use the *recall* measure [14]. Recall captures the fraction of items that have been successfully recommended:

$$recall = \frac{|R_{rel}^q \cap R_{ret}^q|}{|R_{rel}^q|} \tag{2}$$

where R_{ret}^q refers to the relevant items recommended with respect to a query q, and R_{rel}^q refers to all the relevant items with respect to query q.

In order to maximize the *recall*, we argue that *coverage* must be taken into account when building the *U-Net*:

Definition 2 (Coverage). *Given a query q submitted by a user u, TTL and U-Net$_u$ = $\{v_1, ..., v_n\}$, the users in u's U-Net, the coverage is the probability that u can retrieve all items relevant with respect to q.*

However, since a coverage of 1 can be easily achieved by forwarding the queries to all users in the network, we also use the notion of *coverage density*.

Definition 3 (Coverage density). *Given a query q submitted by a user u, TTL and U-Net$_u$ = $\{v_1, ..., v_n\}$, the users in u's U-Net, the coverage density is the contribution to the coverage of each user that receives the query q. It can be expressed in the following way:*

$$coverage\_density(U\text{-}Net_u, q, TTL) = \frac{coverage(U\text{-}Net_u, q, TTL)}{list\_users(U\text{-}Net_u, TTL)} \tag{3}$$

where list_users(U-Net$_u$, TTL) is the set of users that will receive the query q given u's U-Net and TTL.

Intuitively, we want both to maximize the *coverage* and the *coverage density*. Notice that the *recall* is directly related to the *coverage* and is a way to evaluate it.

Problem Definition: Given a user $u \in U$, a query q, I in G, and a gossip based overlay, the goal is to maximize both the *coverage* and the *coverage density* in the following manner:

1. Given two approaches with different coverage (or recall) values, the one with the higher values is preferable;
2. Given two approaches with similar coverage (or recall) values, the one with the highest coverage density is preferable.

3   Diversified Clustering Score

In this section, we show that coverage is important to increase the recall and that usefulness is an excellent way to increase coverage of gossip-based recommendation, and can be used as a clustering score. In the first Sect. 3.1, we develop the coverage probability in the particular case of queries with only one results. Then, in Sect. 3.2, we generalize the probability to the cases where queries can have 1 to n results.

3.1   Development for Queries with One Result

Generally, in state-of-the-art solutions, the *U-Net* of each user is constructed depending of its profile, because we predict that its queries will have a relation with it [2, 4–7].

Here we use the notion of *coverage* and *usefulness*. In other words, we want to maximize the probability that a user u can retrieve the result of any random query q. Therefore, u's neighbors $v_1, ..., v_n \in U$ should be chosen such that the number of relevant items (with respect to the queries u will submit) that can be accessed through them is maximized. However, since we rely on a gossip-based approach where each user computes its *U-Net* individually, we propose the notion of *neighbor's coverage* (*cf.* Definition 4); it is equivalent to the coverage probability presented in Definition 2 when TTL equals 1.

Let Q be the set of all possible queries (all the combinations of terms), $R^q = \{it\}$ the single result of a random query $q \in Q$, V_i^{it} the event "*user v_i shares the item it*" and $\mathcal{P}(V_i^{it})$ its probability. In the following, we first define the neighbor's coverage with respect to *U-Net*$_u = \{v_1, ..., v_n\}$. Then, based on the neighbor's coverage, we express the usefulness of a user v with respect to the other users in u's *U-Net*.

Definition 4 (Neighbor's Coverage). *Given Q and U-Net$_u = \{v_1, ..., v_n\}$, the users in u's U-Net. The neighbor's coverage is the probability that at least one of the user in u's U-Net can return the result $R^q = \{it\}$ of a random query $q \in Q$. The neighbor's coverage is denoted $\mathcal{P}(V_1^{it} \cup V_2^{it} \cup ... \cup V_n^{it})$.*

The user profiles $v_1, ..., v_n$ must be selected such that the neighbor's coverage probability is maximized. Formula 4 develops the neighbor's coverage probability with respect to every user in u's *U-Net*.

$$\mathcal{P}(V_1^{it} \cup ... \cup V_n^{it}) = \sum_{j \in 1,...,n} \left(\mathcal{P}(V_j^{it}) - \mathcal{P}(V_j^{it} \cap (V_1^{it} \cup ... \cup V_{j-1}^{it})) \right) \quad (4)$$

$\mathcal{P}(V_j^{it}) - \mathcal{P}(V_j^{it} \cap (V_1^{it} \cup ... \cup V_{j-1}^{it}))$ represents the coverage added by user v_j with respect to the users $v_1, ..., v_{j-1}$. As a consequence, when $j = 1$, only $\mathcal{P}(V_j^{it})$ is considered as there is no user profiles to compare with.

In the following, we define the usefulness of a user profile v_i with respect to the coverage probability.

Definition 5 (Usefulness). *Given u's U-Net, the usefulness of a user profile v_j is the probability that it can return the result $R^q = \{it\}$ of a random query $q \in Q$, that could not be returned by other users in u's U-Net. In other words, it is defined as follows:*

$$usefulness(v_j | v_{j+1}, ..., v_n) = \mathcal{P}(V_j^{it}) - \mathcal{P}(V_j^{it} \cap (V_1^{it} \cup ... \cup V_{j-1}^{it})s) \quad (5)$$

Formula 5 shows that the usefulness score should consider relevance $\mathcal{P}(V_j^{it})$ and take into account $\mathcal{P}(V_j^{it} \cap (V_1^{it} \cup ... \cup V_{j-1}^{it}))$ which corresponds to the redundancy of user profile v_j with respect to the other user profiles $v_1, ..., v_{j-1}$.

In the following, we show that $usefulness(v_j | v_1, ..., v_{j-1})$ can be expressed into a known probabilistic diversification model [8,9]. In Formula 6 we first factorize usefulness (the right hand side of Formula 5) into a conditional probability.

$$usefulness(v_j | v_{j+1}, ..., v_n) = \mathcal{P}(V_j^{it}) \times (1 - \mathcal{P}(V_1^{it} \cup ... \cup V_{j-1}^{it} | V_j^{it}))$$
$$= \mathcal{P}(V_j^{it}) \times \mathcal{P}(\bar{V}_1^{it} \cap ... \cap \bar{V}_j^{it} | V_j^{it}) \quad (6)$$

Similar to [8,9,15], we assume that the redundancy of a user profile v_1 with another user profile v_2 is independent of its redundancy with other users and we derive Formula 7.

$$\mathcal{P}(V_j^{it}) \times \mathcal{P}(\bar{V}_1^{it} \cap ... \cap \bar{V}_{j-1}^{it} | V_j^{it}) = \mathcal{P}(V_j^{it}) \times \prod_{i \in 1,...,j-1} (1 - \mathcal{P}(V_i^{it} | V_j^{it})) \quad (7)$$

Finally, we observe that the usefulness of a user profile is clearly similar to the probabilistic diversification problem used in [8,9] and presented in Formula 8.

$$usefulness(v_j | v_1, ..., v_{j-1}) = rel(v_j) \times \prod_{i \in 1,...,j-1} (1 - red(v_j, v_i)) \quad (8)$$

where $rel(v_j) = \mathcal{P}(V_j^{it})$ is the relevance of user profile v_j and $red(v_j, v_i) = \mathcal{P}(V_i^{it} | V_j^{it})$ is the redundancy of user profile v_j with respect to the other user profile v_i.

3.2   Development for Queries with m Results

In this section we generalize the probability presented in Eq. 4 for queries with 1 to m results.

We are given $R^q = \{it_1, ..., it_m\}$ the m-results of a random query $q \in Q$. The generalization consists in computing the following probability:

$$\prod_{k \in 1,...,m} \mathcal{P}(V_1^{it_k} \cup ... \cup V_n^{it_k}) =$$

$$\prod_{k \in 1,...,m} \left[\sum_{j \in 1,...,n} \left(\mathcal{P}(V_j^{it_k}) \times \prod_{i \in 1,...,j-1} (1 - \mathcal{P}(V_i^{it_k} | V_j^k)) \right) \right] \quad (9)$$

where $\mathcal{P}(V_j^{it_k})$ is the probability that v_j can return the relevant item $it_k \in R^q$.

In our distributed search and recommendation approach and similarly to the previous section, $rel(v_j) = \mathcal{P}(V_j^{it_k})$ can be computed as the relevance of user v_j with respect to the current user u and $red(v_j, v_i) = P(V_i^{it_k} | V_j^k)$ can be computed as the redundancy of v_j with respect to v_i. Then, since we use $rel(v_j)$ and $red(v_j, v_i)$ to compute Eq. 9, we can see that maximizing the probability of retrieving a single item is the same objective than maximizing the probability of retrieving all items $it_k \in R^q$. In other words, maximizing Eq. 4 would result in a maximization of Eq. 9.

Therefore, in the following, we use the equations presented in Sect. 3.1 because they are simpler.

4 Diversification Algorithms

In the previous section we showed that to increase the neighbor's coverage, diversity must be taken into account. In this section, we propose several approaches to compute a diverse *U-Net*. First in Sect. 4.1 we propose a random-based approach. Then, in Sect. 4.2 we propose an approach that iteratively maximizes the usefulness of the users in *U-Net*. Finally, in Sect. 4.3, we propose an approach that iteratively maximizes the coverage of the *U-Net*.

4.1 Random-Based Diversification

We present in this section a random-based clustering approach. Diversity can easily be achieved using randomness [16]. Indeed, since users have more probabilities to be different than similar, adding randomly the users found on the network in the *U-Net* of a given user u enables to obtain a diverse set of users:

$$score_r(u, v) = rand([0, 1]) \quad (10)$$

Unfortunately these users also have high probability to be different compared to the current user u. Therefore, randomness can be combined with a similarity measure (*e.g. Jaccard*) such as presented in the following equation:

$$score_{rs}(u, v) = sim(u, v) \times rand([0, 1]) \quad (11)$$

Notice that typical similarity-based clustering approaches, where the score of elements in the *U-Net* is unaltered, are no longer suitable when the score is

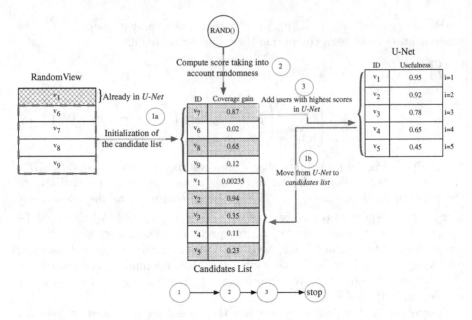

Fig. 2. An example of the execution of random-based *U-Net* clustering.

associated to a random value. When using $score_{rs}(u, v)$, the maximum possible score for user v is $sim(u, v)$. Since, the construction of the *U-Net* is iterative, eventually all users would obtain random values close to 1 and their score would at some point be similar to that obtained with the similarity score. The most similar users would then be promoted to the *U-Net*, and their score frozen. As a result, the *U-Net* will converge to a state containing the same users it would have contained with a score based on similarity, thus ignoring the effect of the randomness. Therefore, we present a modified approach to compute the *U-Net* when using a random-based clustering score.

Based on random gossiping [3], each user u maintains a set of random view entries corresponding to the users profile u is aware of. Periodically, users gossip, and exchange a random subset of their views entries. After the random gossip merging phase, the clustering algorithm is triggered. Then, the algorithm selects the users with the highest scores from the random view and the *U-Net* to build a new *U-Net*. The algorithm uses three main data structures: random view, *U-Net*, and the candidate list. The random view and the *U-Net* are initialized when u joins the network, and continuously updated as a result of random gossip. The candidates list contains the user profiles that will potentially be added to the *U-Net* and is initialized each time the clustering algorithm is triggered. Figure 2 illustrates our approach. The algorithm works in three main steps:

1. The first step consists in initializing the candidates list with the users from both the *Random View* (*i.e.* step 1a) and the *U-Net* (*i.e.* step 1b);

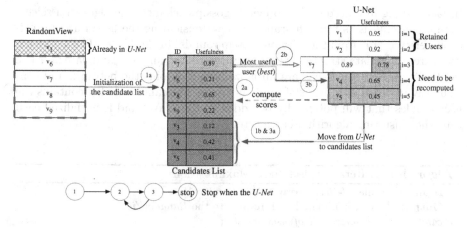

Fig. 3. An example of the execution of *IUM*.

2. The score *sim(u,v)* of each item is computed and associated to a random value;
3. The *N* users from the candidate's list are then inserted in the *U-Net*.

Using randomness to compute a diverse set of users enables to develop very simple clustering approaches such the one presented in Fig. 2 and offers better results than similarity-based approach, as it is described in the experimental evaluation section.

4.2 Iterative Usefulness Maximization (*IUM*)

We now present in details our clustering algorithm that maintains a useful *U-Net* over a random gossip overlay using the usefulness score.

Given the set of users in the random view, the goal of the clustering algorithm is to compute the usefulness of each user found in the view, with respect to those that were previously added to the *U-Net*, taking into account relevance and diversity, as defined in Eq. 8, and to update the *U-Net* as consequence.

Figure 1 shows an example where maximizing iteratively the *usefulness* performs better than using relevance only. Suppose that the relevance metric is the number of items, and the redundancy metric between two users is *Jaccard*. The first user to be selected would be *C* because it shares the maximum number of items and the more relevant and diverse user after *C*. Then the score of *B* is $5 \times \frac{2}{9}$ while the score of *A* is $8 \times \frac{1}{8}$ which is inferior to the score of *B*. In the end, the *usefulness*.

Similarly to the random-based approach, based on random gossiping [3], each user *u* maintains a set of random view entries corresponding to the users profile *u* is aware of. Periodically, users gossip, and exchange a random subset of their views entries. After the random gossip merging phase, the clustering algorithm, which corresponds to the *IUM* Algorithm depicted in Algorithm 1, is triggered.

In fact, taking into account the previous gossip exchange, the algorithm selects the most useful users from the random view considering the useful users previously selected (*i.e.* from the previous gossip rounds) in the *U-Net*. The algorithm uses three main data structures: random view, *U-Net*, and the candidate list. The random view and the *U-Net* are initialized when u joins the network, and continuously updated as a result of random gossip. The candidates list contains the user profiles that will potentially be added to the *U-Net* and is initialized each time the clustering algorithm is triggered.

Algorithm 1. Iterative Usefulness Maximization

Input: u profile, *U-Net$_u$* (array[1..N]), *RandomView$_u$*
Output: *U-Net$_u$* is updated with respect to the *RandomView*

1 *candidates* : *unsorted list of user profiles*;
2 *candidates* $\leftarrow$ *RandomView$_u$* $-$ *U-Net$_u$*;
3 *best* $\leftarrow \emptyset$;
4 $i \leftarrow 0$;
5 **repeat**
6   i++;
7   **for** *each* $c_j \in$ *candidates* **do** $score(c_j) \leftarrow usefulness(c_j, u, U\text{-}Net_u[1..i-1])$;
8   *best* $\leftarrow \arg\max_{c \in candidates}(score(c))$;
9 **until** $i=N$ *or* $score(best) > score(U\text{-}Net[i])$;
10 **if** $score(best) > score(U\text{-}Net[i])$ **then**
11   *after* $\leftarrow$ *U-Net$_u$*$[i..N]$;
12   *U-Net$_u$*$[i] \leftarrow best$ i++;
13   *candidates* $\leftarrow$ *candidates* $-$ *best*;
14   *candidates* $\leftarrow$ *after* $\cup$ *candidates*;
15   *U-Net$_u$* $\leftarrow$ *U-Net$_u$* $-$ *after*;
16   **while** $i < N$ **and** *candidates*$\neq \emptyset$ **do**
17   **for** *each* $c_j \in$ *candidates* **do**
18          $score(c_j) \leftarrow usefulness(c_j, u, U\text{-}Net_u[1..i-1])$;
19   *best* $\leftarrow \arg\max_{c \in candidates}(score(c))$;
20   *U-Net$_u$*$[i] \leftarrow best$;
21   *candidates* $\leftarrow$ *candidates* $-$ *best*;
22   i++;

In the following we present in more details the *IUM* algorithm based on the example of Fig. 3. The random view entries correspond to the profiles of users v_1, v_6, v_7, v_8, v_9. The previous useful user profiles are v_1, v_2, v_3, v_4, v_5 and are stored in *U-Net*. Assuming that the algorithm is executed in u's node, the algorithm input is u's profile, its random view denoted *RandomView$_u$* and its *U-Net* denoted *U-Net$_u$*. The data structure used for *U-Net* is an array of size N of user profiles, associated to their usefulness score and sorted in decreasing order of usefulness. The output of the algorithm is the updated *U-Net*. *IUM* algorithm has three main parts:

1. The first part (lines 1 to 9) finds the *best* useful user profile from the random view, and the position i where it should be inserted in the *U-Net* (recall that the usefulness score of a user depends on its position in the *U-Net*). As a consequence, the update of the *U-Net* will only concern the user profiles from position i to N. To find the *best* useful user from the random view, the algorithm first initializes the *candidates* list with all users in the random view except those already in the *U-Net* (line 2). In Fig. 3, v_1 is already in the *U-Net*, so the candidates list is initialized with the users v_6, v_7, v_8, v_9 (1a). For each position i in *U-Net*, all the usefulness scores of the candidates are computed using Formula 8 taking into account the set of users in the *U-Net* at positions $1, ..., i - 1$, and compared with the usefulness score of the user profile in $U\text{-}Net_u[i]$. If the best user profile in *candidates* is more useful than $U\text{-}Net_u[i]$, then, the algorithm stops iterating (line 9). If there is more than one best user profile, the best user profile is chosen randomly with respect to the set of best user profiles. In Fig. 3, v_7 is more useful than v_3 at the third position in u's *U-Net* because v_3's usefulness is 0.78 while v_7's usefulness is 0.89 (1b). If there is no user profile in the candidate list whose profile score is superior to any user profile in the *U-Net*, position N is reached and the algorithm stops. Only the scores of the user profiles up to position i are definitive. Thus, in our example, the scores of v_4, v_5, v_6, v_8, v_9 are not definitive because they are either not in the *U-Net* or after i.
2. The second part (lines 10 to 15) copies and deletes the remaining user profiles (from position i to N) from the *U-Net* to the candidates (2a) list because their scores need to be recomputed using Formula 8 and with respect to the best user profile in *candidates* (computed in part 1). Then, the best user profile is inserted in position i. In the on-going example of Fig. 3, the user profiles v_3, v_4, v_5 are copied and removed from the *U-Net* to the candidates list and user profile v_7 is added in the *U-Net* at position 3 (2a and 2b).
3. Finally, in the last part (lines 16 to 22), the algorithm iteratively computes, for each empty position i in the *U-Net* (positions emptied in part 2), the scores of the user profiles in the candidates list using Formula 8 and taking into account the set of users in the *U-Net* at positions $1, ..., i - 1$ (lines 17 and 18 and step 3a in the figure). Then, the most useful candidate is moved to the *U-Net* at that position (line 20 and step 3b in the figure). The algorithm repeats these steps until all the positions in *U-Net* are filled out (line 16).

Recall that gossip protocols converge quickly [4]. As a consequence the *U-Net* will also converge quickly and, in general, tends to stabilize. Therefore, the algorithm will stop at step 1b more and more frequently.

4.3   Iterative Coverage Maximization (*ICM*)

In this section we present an approach that iteratively maximizes the coverage. First, we show that maximizing the coverage of the *U-Net* and maximizing the usefulness of the users in the *U-Net* (*cf.* previous section) are two different things. Then, we present our approach to maximize the coverage. Finally, we discuss the properties of our heuristics.

Coverage Maximization vs Usefulness Maximization: We explain the difference through the example shown in Fig. 1. We are given a *U-Net* composed of two users A and B. Suppose that we iteratively maximize the *usefulness* of the users in the *U-Net* such as presented in the previous section. After a gossip round, the clustering algorithm is triggered and a user C is in the *RandomView*. We can observe in Fig. 1 the following characteristics:

$$coverage(C) > coverage(A)$$
$$coverage(C) > coverage(B) \tag{12}$$

Since we maximize the *usefulness*, after the clustering step, the user $A \in$ *U-Net* will be replaced by the user C. However, as it can be observed in Fig. 1 the coverage of $C \cup A$ or $C \cup B$ is lower than that of $A \cup B$

$$A \cup B > C \cup B$$
$$> C \cup A \tag{13}$$

Therefore, maximizing iteratively the *usefulness* of the users in the *U-Net* may result in a reduction of the coverage, which is not the case when maximizing iteratively the *coverage*.

Iterative Coverage Maximization Approach: Given the set of users in the random view, the goal of the clustering algorithm is to compute the coverage gain of each user found in the view, with respect to those that were previously added to the *U-Net*, taking into account relevance and diversity, as defined in Eq. 8, and to update the *U-Net* in consequence.

Similarly to the random-based and *IUM* approaches, based on random gossiping [3], each user u maintains a set of random view entries corresponding to the user profiles u is aware of. Periodically, users gossip, and exchange a random subset of their views entries. After the random gossip merging phase, the clustering algorithm, which corresponds to the *ICM* Algorithm depicted in Algorithm 2, is triggered. In fact, taking into account the previous gossip exchange, the algorithm selects users from the random view with the highest coverage gain considering the useful users – recall that the coverage of the usefulness of the users in *U-Net* – previously selected (*i.e.* from the previous gossip rounds) in the *U-Net*. The algorithm uses three main data structures: random view, *U-Net*, and the candidate list. The random view and the *U-Net* are initialized when u joins the network, and continuously updated as a result of random gossip. The candidates list contains the user profiles that will potentially be added to the *U-Net* and is initialized each time the clustering algorithm is triggered. In the following we present in more details the *ICM* algorithm based on the example of Fig. 4. The random view entries correspond to the profiles of users v_1, v_6, v_7, v_8, v_9. The previous useful user profiles are v_1, v_2, v_3, v_4, v_5 and are stored in *U-Net*. Assuming that the algorithm is executed in u's node, the algorithm input is u's profile, its random view denoted *RandomView$_u$* and its *U-Net* denoted *U-Net$_u$*. The data structure used for *U-Net* is an array of size N of user profiles, associated to their usefulness score.

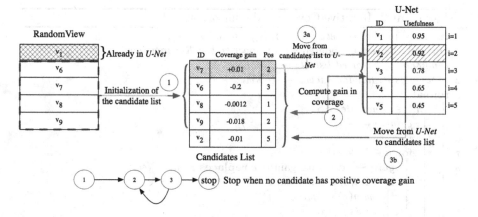

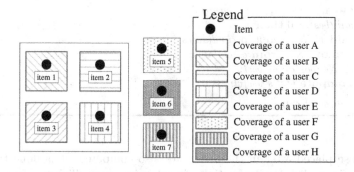

Fig. 4. An example of the execution of Max-coverage-Unet.

Fig. 5. Limit of the Iterative Coverage Maximization Heuristic.

The output of the algorithm is the updated *U-Net*. *Iterative Coverage Maximization* algorithm has three main parts:

1. The first step (lines 1 and 2) consists in initializing the candidate's list with all users from the *random view* except those who already are in the *U-Net*.
2. The second step (lines 4 to 12) consists in computing the maximum coverage gain of all candidates. For each candidate, the gain is the maximum gain the candidate could provide to the coverage by replacing a user in the *U-Net*. Once this maximum gain is found, the candidate is associated to the both the gain's value, and the indice of the user in the *U-Net* it should replace to effectively affect the coverage. In Fig. 4, the user v_7 obtains a positive gain of 0.01 by replacing v_2 at position 2 in the *U-Net*. Notice that all other candidates are assigned negative gain values, meaning that they would reduce the coverage of the *U-Net*.
3. The last step (lines 13 to 18) consists in finding the *best* candidate (the one with the highest gain) and exchanging it in with the corresponding user in the *U-Net* if the gain is positive. The *usefulness* of the best candidate, along

Algorithm 2. Iterative Coverage Maximization

Input: u profile, $U\text{-}Net_u$ (array[1..N]), $RandomView_u$
Output: $U\text{-}Net_u$ is updated with respect to the $RandomView$
1 $candidates$: *unsorted list of user profiles*;
2 $candidates \leftarrow RandomView_u - U\text{-}Net_u$;
3 **repeat**
4   **for** $c \in candidates$ **do**
5        $gain(c) \leftarrow -\infty$;
6        $i \leftarrow 1$;
7   **for** $u \in U\text{-}Net$ **do**
8           $gain \leftarrow$ **compute** gain if c replaces u in $U\text{-}Net_u$;
9   **if** $gain(c) < gain$ **then**
10             $gain(c) \leftarrow gain$;
11             $pos(c) \leftarrow i$;
12          $i++$;
13    $best \leftarrow \arg\max_{c \in candidates}(gain(c))$;
14   **if** $gain(best) > 0$ **then**
15   **add** $U\text{-}Net[pos(c)]$ **in** $candidates$;
16   **remove** $best$ **from** $candidates$;
17       $U\text{-}Net[pos(c)] \leftarrow best$;
18   **compute** $best$'s usefulness;
19 **until** $gain(best) \leq 0$;

---

wit those placed after in the array, are also computed. It is done this way because it renders the computing of next gains more efficient. Finally, if an exchange has been produced, the algorithm comes back to the second step; otherwise, the algorithm stops.

This algorithm is an heuristic and does not necessarily converge towards the maximum possible coverage. However, it converges towards a local maximum and therefore it eventually guarantees that:

$$\nexists u \in U, u \notin U\text{-}Net, v \in U\text{-}Net \mid coverage(U\text{-}Net \cup u \setminus v) > coverage(U\text{-}Net) \quad (14)$$

In other words, the *U-Net* will converge to a state where it will not be possible to find a user that is not in the *U-Net* that can increase its coverage by an individual exchange.

Discussion About the Heuristic: Figure 5 shows an example where, contrary to Fig. 4, it is better to use *IUM* than *ICM*.

Example 2. Suppose that at the beginning, the *U-Net* is composed of users B, C, D and E and that the relevance metric is the number of items shared by the user. The current *U-Net* enables to retrieve 4 items. Replacing a user from the *U-Net* with another that is not yet in the *U-Net* does not enable to increase the number of retrievable items (*i.e.* coverage). Therefore, given these users, we can say that the *U-Net* has converged with respect to the *ICM* approach.

By using the *Iterative Usefulness Maximization* algorithm, the *U-Net* would have been composed at first of the user A because A shares 4 items, followed by F, G and H, thus enabling to retrieve 7 items.

This low quality of *ICM* is due to the fact that our approach is an heuristic and does not compute the complete set of combination for the *U-Net* to select the maximized coverage which is impractical to do because of the complexity of such an approach. In other words, *ICM* does not guarantee that:

$$\nexists U\text{-}Net', U\text{-}Net' \neq U\text{-}Net \mid coverage(U\text{-}Net') > coverage(U\text{-}Net) \qquad (15)$$

The *ICM* heuristic has a very small average complexity compared to an approach that would compute all combinations. For instance, in our experiments on the small dataset of *MovieLens*, with a random view of size 20 and a *U-Net* of size 8, in the worst case, the average number of iterations is 325, the worst number of iterations for a specific user is 800 and it converges through time towards a value of $160 = m \times N$ where m is the size of the random view and N is the size of the *U-Net*. An approach computing all combinations would generate $(m+N)!/(m! \times N!)$ of them, knowing that for each of them, the *usefulness* of all user in the *U-Net* needs to be computed. With $m = 20$ and $N = 8$, the number of different *U-Net* to compute is $24,864,840$ – although optimizing strategies such as pruning could be possible, the complexity would still be too high.

5   Experimental Evaluation

In this section, we provide an experimental evaluation to validate our approach and compare it to other state-of-the-art solutions. We conducted a set of experiments using four datasets which correspond to *MovieLens-small*[1], *MovieLens*[2], *LastFM*[3] and *Delicious*[4]. In Sect. 5.1, we introduce the experimental setup of our evaluation. In Sect. 5.2, we discuss the characteristics of our four datasets. Then, in Sect. 5.3, we present and discuss the experimental results. Finally, in Sect. 5.4 we propose a general discussion of our results.

5.1   Experimental Setup

We performed our experiments through simulation with real data. We used four different datasets: *MovieLens-small*, *MovieLens*, *LastFM* and *Delicious*. *MovieLens-small* and *MovieLens* datasets are composed of users that rated movies. *LastFM* dataset is composed of users that listened to artists. Each user

---

[1] http://grouplens.org/datasets/movielens/.
[2] http://grouplens.org/datasets/movielens/.
[3] http://grouplens.org/datasets/hetrec-2011/.
[4] http://grouplens.org/datasets/hetrec-2011/.

also associated tags to the artists he listened to. Finally, *Delicious* dataset is composed of users that shared and associated tags to bookmarks.

The queries used in the experiments consist of:

- In *MovieLens-small* and *MovieLens*, for each user, a random subset of movies are shared and the rest are used as the queries to submit. In particular, the words in the title are used as separate keywords.
- In *lastFM* and *Delicious* queries are computed as the random association of several tags submitted by a given user on a given item.

An experiment is composed of two parts. First, all users gossip during 400 rounds until convergence. Then, every 20 gossip rounds all users submit one of their queries. The experiment stops at 500 gossip rounds. We measure the average recall results. The recall enables to compute the fraction of items that has been successfully recommended as presented in Sect. 2. On the *MovieLens* dataset, the recall value is 1 if the movie been found and 0 otherwise. On *LastFM* and *Delicious*, the recall is the proportion of pictures in the whole dataset that contains all query's keywords that have been returned to the user. More generally, the recall is evaluated as follows:

$$recall = \frac{|R_{rel}^q \cap R_{ret}^q|}{|R_{rel}^q|} \tag{16}$$

where R_{rel}^q is the set of relevant items with respect to the query q and R_{ret}^q the set of items that have been effectively retrieved during the processing of q.

In our experiments, we used the relevance metrics presented in Table 1; however, since the results are very similar between these metrics, we only show the results for the *Jaccard* metric[5]. Indeed, the better a similarity measure is, the higher the risk of redundancy in a cluster and the higher the gain of diversity in terms of recall.

The following approaches have been compared:

1. **Similar (s):** given a user u, its most similar users are added in $U\text{-}Net_u$;
2. **Similar and Random (rs):** the approach presented in Sect. 4.1;
3. **Iterative Usefulness Maximization or *IUM* (u):** the approach presented in Sect. 4.2;
4. **Random (r):** the users are randomly added to the *U-Net*;
5. **Iterative Coverage Maximization *ICM* (c):** the approach presented in Sect. 4.3.

In our experiments, the size of the random view has been fixed to 20 but it is not important as it only modifies the convergence speed.

Notice that, although other complementary contributions (e.g., replication, different aggregation protocols) might also be used, their recall is highly dependent on the underlying similarity metric employed. Thus, our results, which focus on the improvements of the similarity measures, are translatable to such techniques.

---

[5] Similar results are expected for other techniques not tested such as Dice [11], Tanimoto [12].

5.2   Datasets

Each dataset has different features, in particular users are more or less redundant if both the number of items per user and the total number of items is more or less respectively. The characteristics of the datasets are summarized in the following table (Table 2).

Table 2. Datasets characteristics.

Dataset	Items	# items	# users	Avg items/user
MovieLens-small	Movies	1,700	1,000	100
MovieLens	Movies	3,900	6,040	166
LastFM	Artists	16,543	1,892	25
Delicious	Bookmarks	46,876	2,000	109

Additionally, Fig. 6 presents the distribution of the queries' results among the datasets. Given a dataset, we have performed a hierarchical clustering analysis on the users and sorted them according to the resulting tree. We use this ordering to represent the users in the figures, so that similar users are close to each other on the axis. The abscissa represents the users submitting their queries while the ordinate corresponds to users that share items relevant with respect to those queries. Each cell is a group of 50 users and the cell's color corresponds to the number of items (*i.e.* ordinate) relevant with respect to the corresponding users' queries (*i.e.* abscissa). We now discuss the datasets properties.

Movielens-Small and Movielens: we can observe that in both datasets there is no correlation between the profiles of the users submitting the queries and those sharing their results – a correlation would be characterized by a visible diagonal. However, we can identify special types of users depending on their behavior, namely, users which share a lot of popular items (dark rows) and users that search popular items (dark columns).

LastFM: in this dataset,the correlation between the profile of the user submitting queries and those sharing the results is weak. We can observe, though, that around 25 % of users (from $1,300$ to $1,900$) share less number of items and in those cases the correlation is slightly higher.

Delicious: a strong correlation between the users submitting the queries and those sharing their results can be observed, as indicated by the dense diagonal in the figure.

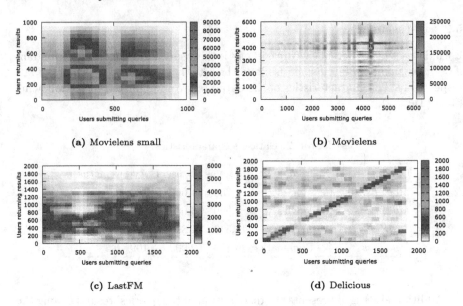

(a) Movielens small   (b) Movielens

(c) LastFM   (d) Delicious

Fig. 6. Users submitting queries vs users sharing the results density.

5.3   Experiments

We divide the experiments into two sets. In the first set, with $TTL=1$, the number of queried users is the same for all strategies. In this case, the *neighbor's coverage* (*cf.* Definition 4 in Sect. 3) is also an exact model for the query. With $TTL > 1$, the *U-Nets* of the queried users forward the query subsequently, so the probability model is not exact, although a good approximation as seen in the results. Moreover, different strategies may involve a different number of users, so other concepts such as recall density are explored.

TTL=1: Figure 7 presents the recall results when $TTL=1$ for various sizes of *U-Net*. We can observe that for diversified approaches (*i.e.* **rs, u, c**) the recall results are often the best, with u (*i.e. IUM*) being generally the best approach.

In Fig. 7a and b, the random-based approach performs well while the similarity-based approach does not. This can be explained with Fig. 6a and b where we can see that there is no evident correlation between the profiles of the users submitting the queries and those sharing the results.

In Fig. 7c, on the other hand, the random-based approach performs the worst. This can be explained with the dataset characteristics. First, a small correlation exists between the profiles of the users submitting the queries and those sharing the results. Second, around 25 % of the users do not share relevant results except for users whose profile is similar. Thus, selecting the users randomly to build the *U-Net* is not a good approach, since those users will be selected with the same probability as the others.

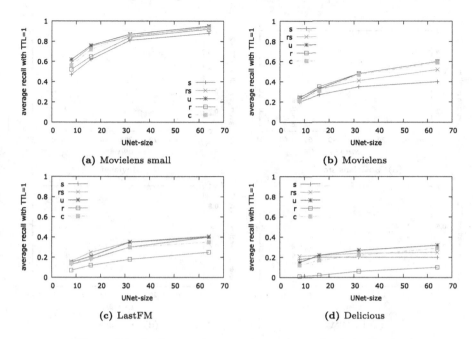

Fig. 7. Recall when $TTL = 1$ with respect to the *U-Net*'s size.

Finally, in Fig. 7d, the random-based approach also performs the worst. This can be explained with the datasets characteristics showed in Fig. 6d. Since a clear correlation exists between the users submitting the queries and those sharing the results, it is better to chose similar users when building *U-Net* rather than random ones. Additionally, in this same figure, we can observe that increasing the *U-Net* size does not enable to increase the similarity-based approach recall. This can be explained with the *Profile Redundancy* effect: the users in *U-Net* share the same items because they are redundant.

TTL>1: Figure 8 shows the recall results when $TTL = 2$. We obtain similar results compared to those obtained with $TTL=1$. The only exception is the gain observed with the random-based approach. Figure 9 can explain this gain. Since all solutions except the random-based one exploit a similarity score, there is a high probability for the query to be forwarded to the same users several times, thus reducing the total number of users involved in query processing – this is the *Network Redundancy* effect. In all datasets, the random-based approach involves a much higher number of users than the other approaches while recall results are similar to the other approaches. This holds even in the two datasets where there is a correlation between the profiles of the users submitting the queries and those sharing the results.

Figure 10 combines the two previous figures to compute the *recall density*. Although the similarity-based approach obtain low recall results, since the number of users involved during query processing is even lower, the *recall density* is

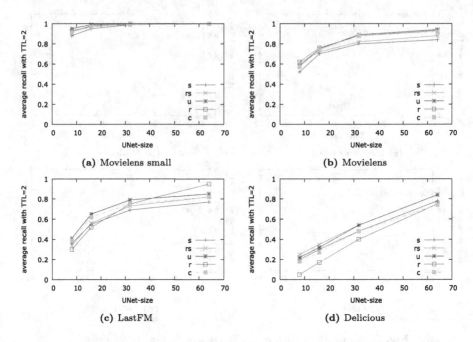

Fig. 8. Recall when $TTL = 2$ with respect to the *U-Net*'s size.

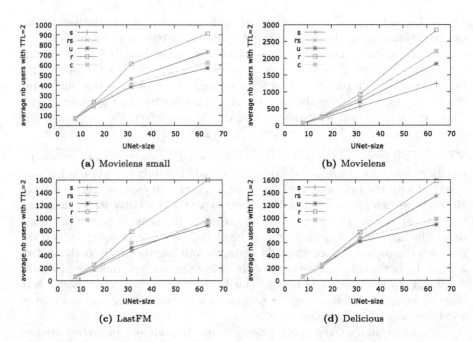

Fig. 9. Number of users involved during query processing when $TTL = 2$ with respect to the *U-Net*'s size.

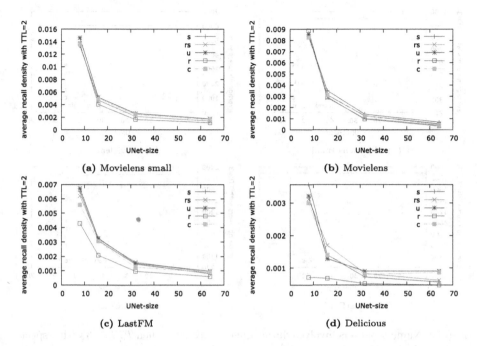

Fig. 10. Recall density when $TTL = 2$ with respect to the *U-Net*'s size.

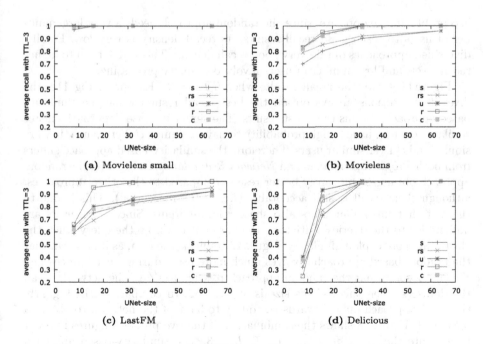

Fig. 11. Recall density when $TTL = 3$ with respect to the *U-Net*'s size.

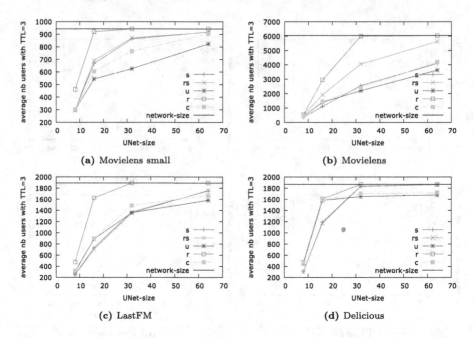

Fig. 12. Number of users involved during query processing when $TTL = 3$ with respect to the *U-Net*'s size.

acceptable. In the opposite, since the random approach needs a very large number of users to reach good recall results, its recall density is very low. Finally, diversified approaches obtain very good recall density. This is due both to higher recall levels and low number of users involved in query processing.

Figure 11 shows the recall results when $TTL = 3$. Except in Fig. 11c, the diversified solutions always obtain the best recall results while the similarity based approach obtains the worst results. The similarity-based approach cluster similar users who have a high probability to have a similar *U-Net* – since they are similar and cluster similar users. Therefore, the similarity-based approach suffers from both *Profile Redundancy* and *Network Redundancy*. In Fig. 11c, the random approach obtains significantly better results compared to the other approaches, although their recall is still acceptable (*i.e.* > 0.8 in general). This is due to the fact that un-similar users also share relevant items. Since these users are not similar to the queries' initiators, they are not reach by the query when the clustering score exploit similarity. Figure 12 confirms the idea, as it is shown that the random-based approach involves much more users during query processing. We can observe that the random approach reaches 100 % of the network on all the datasets when the *U-Net*'s size is at least 32. In comparison, In Fig. 11b, the *IUM* approach only forwards the query to 33 % of the network to obtain a recall of 1. Figure 13 shows the combination of the two previous figures in order to evaluate the *recall density* when $TTL = 3$. The similarity-based approach sometimes obtains high *recall density* but since it corresponds to very low recall

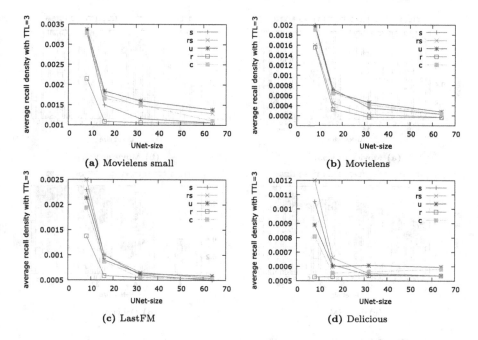

Fig. 13. Recall density when $TTL = 3$ with respect to the *U-Net*'s size.

results compared to the other solutions it is not a good result. In contrary, the diversified solutions, which had very good recall results and low number of users, obtain a very good recall density. The random approach needs to forward the query to a large number of users to obtain good recall results, and therefore, has a low recall density.

5.4   General Discussion

Several aspects have been pointed out in the previous experiments that we are going to discuss here. Figure 14 is a synthesis of all previous experiments on recall. Two main elements can be observed. First, when a correlation exists between the users submitting the queries and those sharing their results, all methods except the random one are better when $TTL = 1$. Second, in all cases, when $TTL > 1$, the random-based approach, by forwarding the query to more users than the other approaches obtain good recall results.

In Table 3, we represent the previous results in relative terms. More precisely, we have normalised the values for all experiments with respect to the best approach – given a TTL, a *U-Net*'s size, we divide the recall of all approaches by the highest recall obtained in the corresponding experiment. The numbers in bold and green correspond to the two best values observed in a column and those underlined in red correspond to the two worst results. High recall results are good while a low number of users is good for scalability. Table 4 represents an

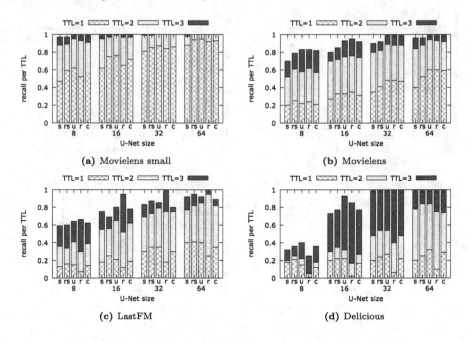

Fig. 14. Recall gain per TTL with respect to the U-Net's size.

Table 3. Normalized average recall and average number of users.

Approach	TTL=1		TTL=2		TTL=3		Average	
	Recall	# Users	Recall	# Users	Recall	# Users	Recall	# Users
Similar	83%	100%	90%	**76%**	92%	**71%**	88%	**82%**
RS	96%	100%	**94%**	82%	92%	77%	**94%**	86%
IUM	**97%**	100%	**97%**	79%	**96%**	72%	**97%**	83%
Random	52%	100%	81%	100%	**93%**	100%	75%	100%
ICM	**85%**	100%	91%	83%	**93%**	76%	90%	86%

analysis of the approaches in terms of *relevance, profile redundancy* and *network redundancy*.

The similarity-based approach and the random-based approach show to have the worst results in terms of recall in average. These results can be explained with Table 4. Although the similarity-based approach is good in terms of relevance (*i.e.* the *U-Net* is filled with relevant users), it suffers from both profile and network redundancy. The profile redundancy of similarity is confirmed with the low recall values when $TTL = 1$ while the network redundancy effect manifest when $TTL > 1$. Although the random approach does not suffer from redundancy, it suffers from low relevancy. This explains why the approach obtains lower recalls when TTL is 1 and 2. When $TTL=3$, due to its low network redundancy, it compensates the relevance problems by reaching a higher number of user, which, nonetheless, is bad in terms of scalability.

Table 4. Approaches analysis.

Approach	Relevance	Profile redundancy	Network redundancy
Similar	High	High	Medium
Random Similar	Medium	Medium	Medium
IUM	High	Low	Medium
Random	Low	Low	Low
ICM	High	Low	Medium

IUM and *ICM* obtain the best results in terms of recall. Additionally, *IUM* attain very good results in terms of the number of users involved in query processing. *ICM*, on the other hand, gets an average behavior, with results worse than *IUM* in all cases, but not as bad as random. Random-similar is also a good compromise between random and similar. It obtains better recall than any of them while accessing much less users than random.

Figure 15 represents the same results of previous experiments, but with another perspective. It has to be read in the following way: given a recall value, what is the number of users involved during query processing for a specific approach.

For instance, in Fig. 15b, if we want a recall of 1, we need 3 times less users with *IUM* than random and 2.74 less for the *ICM* approach. It is impossible to reach a recall of 1 with the similarity-based and random-similar approaches.

In Fig. 15d, if we want a recall of at least 0.85, *IUM* needs 860 users, random-similar needs 1,300, both *ICM* and random needs 1,600 and the similarity based approach needs 1,800.

More generally, we can observe that the diversified approaches and specifically the *IUM* approach obtain the best results in all datasets (which would correspond with the skyline), while the worst results are obtained by the similarity-based approach in Figs. 15a and b and the random in Fig. 15c and d.

6 Related Work

Distributed recommendation for web data based on collaborative filtering has been recently proposed with promising results. In this section, we compare our recommendation approaches with state of the art solutions.

In [17], Loupasakis and Ntarmos propose a decentralized approach for social networking with three goals in mind: privacy, scalability with profitability and availability. They propose an architecture based on a DHT for keywords query search. Since, DHTs are better suited for exact-match queries, the authors propose to decompose each query into several single word exact-match queries. The main drawback is that responses that have medium scores with respect to each keyword but high scores with respect to all the keywords are likely to be missed.

P2PRec [4] is a gossip-based search and recommendation solution where the profile of each user u is represented as a set of topics computed based u's items.

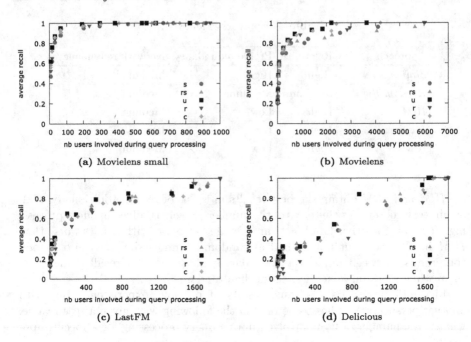

Fig. 15. Experiments synthesis.

Then, using gossip protocols, similar users in terms of topics, are clustered together and used to guide recommendation as we do. However, since diversity is not taken into account, users within each cluster can be redundant, thus limiting recall results.

TRIBLER [18] is also a gossip-based approach for search and recommendation where user profiles are computed as the set of items shared by the current users. Then, using gossip protocoles, similar users are clustered together. Additionally, friendship can be taken into account. The similar users and the friends are used to guide recommendation. However, the lack of diversity results in redundant users in the cluster.

In [6], Kermarrec *et al.* focus on recommendation and propose to combine gossip algorithms and random walks. The users are clustered based on relevance through gossip protocols. A user has knowledge of the items shared by its neighbors. To compute the recommendation, each user runs locally a random walk using a transition similarity matrix. However, the computational complexity of the algorithm with respect to the size of the neighborhood and the number of items. Reduces the scalability of the approach.

Moreover, Kermarrec *et al.* [7] claim that, since users are heterogeneous, the similarity measure used to cluster users should also be heterogeneous. Nevertheless, the concept of diversity is different from ours as it represents the usage of various relevance scores depending on each user's profile. As a consequence, each user's cluster may still carry redundant user profiles, because there is no explicit diversification.

In [1], Bai *et al.* propose a solution for personalized *P2P top-k search* in the context of collaborative tagging systems, called *P4Q*. In this solution, the users are clustered based on relevance through gossip protocols. The users in each cluster are split into two groups: (1) the c closest users from which u replicates all items metadata (*i.e.* tagging actions) and (2) the n less similar users from which u knows only the profile. Still, diversity is not taken into account and users within the clusters are likely to be redundant.

In [19], the authors focus on the idea of *semantic peers*. Contrary to us, they use a *Semantic Overlay Network* to group peers as we did with our random view. Then, they propose that each peer chose its most similar acquaintances in a light weight structure similar to our *U-Net*. Thus, similarly to the other work, they do not benefit from diversification and their solution may have low recall in some scenarios.

In [20], the authors propose a solution for recommendation in gossip-based *P2P* approaches. Similarly to our scenario, they argue that *P2P* is an interesting way to diffuse the information among dynamic communities. However, they also gather peers using a similarity metric, thus loosing the novelty (*i.e.* new items) that could provide diversification as we did.

7   Conclusion

In this paper, we proposed a new gossip-based search and recommendation approach with new measures and techniques.

We first showed that diversity, is very effective in increasing recall results and we proposed several new clustering scores to compute a diversified *U-Net*. Then, we designed new clustering algorithms with respect to our diversity-based scores.

We validated our proposal with an experimental evaluation using several datasets and show major gains with recall results more than 70 % better than similarity-based approaches and up to 22 times better than random based approaches. Additionally while reaching very high recall results, diversity-based approaches only involve a small number of users during query processing compared to other solutions.

Acknowledgments. Experiments presented in this paper were carried out using the Grid'5000 testbed, supported by a scientific interest group hosted by Inria and including CNRS, RENATER and several Universities as well as other organizations (see https://www.grid5000.fr).

References

1. Bai, X., Guerraoui, R., Kermarrec, A., Leroy, V.: Collaborative personalized top-k processing. Trans. Database Syst. **36**(26), 1–38 (2011)
2. Carretero, J., Isaila, F., Kermarrec, A.M., Taïani, F., Tirado, J.M.: Geology: modular georecommendation in gossip-based social networks. In: ICDCS, pp. 637–646 (2012)

3. Jelasity, M., Babaoglu, O.: T-Man: gossip-based overlay topology management. Engi. Self-Org. Syst. **53**(13), 1–15 (2006)
4. Draidi, F., Pacitti, E., Parigot, D., Verger, G.: P2Prec: a social-based P2P recommendation system. In: CIKM, pp. 2593–2596 (2011)
5. Voulgaris, S., van Steen, M.: Epidemic-style management of semantic overlays for content-based searching. In: Cunha, J.C., Medeiros, P.D. (eds.) Euro-Par 2005. LNCS, vol. 3648, pp. 1143–1152. Springer, Heidelberg (2005)
6. Kermarrec, A.-M., Leroy, V., Moin, A., Thraves, C.: Application of random walks to decentralized recommender systems. In: Lu, C., Masuzawa, T., Mosbah, M. (eds.) OPODIS 2010. LNCS, vol. 6490, pp. 48–63. Springer, Heidelberg (2010)
7. Kermarrec, A.M., Taïani, F.: Diverging towards the common good: heterogeneous self-organisation in decentralised recommenders. In: SNS, pp. 3–8 (2012)
8. Angel, A., Koudas, N.: Efficient diversity-aware search. In: SIGMOD, pp. 781–792 (2011)
9. Chen, H., Karger, D.R.: Less is more: probabilistic models for retrieving fewer relevant documents. In: SIGIR, pp. 429–436 (2006)
10. Servajean, M., Pacitti, E., Liroz-Gistau, M., Amer-Yahia, S., El Abbadi, A.: Exploiting diversification in gossip-based recommendation. In: Hameurlain, A., Dang, T.K., Morvan, F. (eds.) Globe 2014. LNCS, vol. 8648, pp. 25–36. Springer, Heidelberg (2014)
11. Dice, L.R.: Measures of the amount of ecologic association between species. Ecology **26**(3), 297–302 (1945)
12. Rogers, D.J., Tanimoto, T.T.: A computer program for classifying plants. Science **132**(3434), 1115–1118 (1960)
13. Manning, C.D., Raghavan, P., Schütze, H.: Introduction to Information Retrieval. Cambridge University Press, Cambridge (2008)
14. Baeza-Yates, R., Ribeiro-Neto, B.: Modern Information Retrieval. ACM Press, New York (1999)
15. Carbonell, J., Goldstein, J.: The use of MMR, diversity-based reranking for reordering documents and producing summaries. In: SIGIR, pp. 335–336. ACM Press, New York (1998)
16. Anagnostopoulos, A., Broder, A., Carmel, D.: Sampling search-engine results. In: WWW 2005 (2005)
17. Loupasakis, A., Ntarmos, N.: eXO: decentralized autonomous scalable social networking. In: CIDR, pp. 85–95 (2011)
18. Pouwelse, J., Garbacki, P., Wang, J., Bakker, A., Yang, J., Iosup, A., Epema, D.H.J., Reinders, M., Van Steen, M.R., Sips, H.J.: TRIBLER: a social-based peer-to-peer system. Concurr. Comput. Pract. Exp. **20**(2), 127–138 (2008)
19. Penzo, W., Lodi, S., Mandreoli, F., Martoglia, R., Sassatelli, S.: Semantic peer, here are the neighbors you want! In: Proceedings of the 11th International Conference on Extending Database Technology: Advances in Database Technology, EDBT 2008, pp. 26–37. ACM, New York (2008)
20. Baraglia, R., Dazzi, P., Mordacchini, M., Ricci, L.: A peer-to-peer recommender system for self-emerging user communities based on gossip overlays. J. Comput. Syst. Sci. **79**(2), 291–308 (2013)

Hypothesis Discovery Exploiting
Closed Chains of Relations

Kazuhiro Seki$^{(\boxtimes)}$

Konan University, Kobe, Hyogo 658-8501, Japan
seki@konan-u.ac.jp
http://www.konan-u.ac.jp/hp/seki/

Abstract. The ever-growing literature in biomedicine makes it virtually impossible for individuals to grasp all the information relevant to their interests. Since even experts' knowledge is limited, important associations among key biomedical concepts may remain unnoticed in the flood of information. Discovering those hidden associations is called hypothesis discovery or literature-based discovery. This paper propose an approach to this problem taking advantage of a closed, triangular chain of relations extracted from the existing literature. We consider such chains of relations as implicit rules to generate explanatory hypotheses. The hypotheses generated from the implicit rules are then compared with newer knowledge for assessing their validity and, if validated, they are served as positive examples for learning a regression model to rank hypotheses. As a proof of concept, the proposed framework is empirically evaluated on real-world knowledge extracted from the biomedical literature. The results demonstrate that the framework is able to produce legitimate hypotheses and that the proposed ranking approach is more effective than the previous work.

Keywords: Literature-based discovery · Text mining · Inference · Semi-supervised learning · Hypothesis ranking

1 Introduction

The amount of scientific knowledge is rapidly growing beyond the pace one could digest. For example, Medline[1], the most comprehensive bibliographic database in life science, currently contains over 19 million references to journal articles and 2,000–4,000 completed references are added each day. Given the substantial volume of the publications, it is virtually impossible for individuals to deal with the information without the aid of intelligent information processing techniques, such as information extraction [6] and text data mining (TDM) [2,19].

TDM aims to discover heretofore unknown knowledge through an automatic analysis on textual data. A pioneering work in TDM, also known as literature-based discovery or hypothesis discovery, was conducted by Swanson in the 1980's.

---

[1] http://www.ncbi.nlm.nih.gov/entrez.

© Springer-Verlag Berlin Heidelberg 2015
A. Hameurlain et al. (Eds.): TLDKS XXII, LNCS 9430, pp. 145–164, 2015.
DOI: 10.1007/978-3-662-48567-5_5

He argued that there were two premises logically connected but the connection had been unnoticed due to overwhelming publications and/or over-specialization. To demonstrate the validity of the idea, he manually analyzed a number of articles and identified logical connections implying a hypothesis that fish oil was effective for clinical treatment of Raynaud's disease [31]. The hypothesis was later supported by experimental evidence [11].

This study is motivated by the series of Swanson's work [32–35] and attempts to advance the research in hypothesis discovery. Specifically, we aim to address two problems that the existing work has generally suffered from. One is the unknown nature of a generated hypothesis. Traditional co-occurrence-based approaches only identify two potentially associated concepts, leaving the meaning of the association unknown, which requires experts to interpret the hypothesis. This vagueness has significantly limited the utility of hypothesis discovery. To cope with the problem, we derive hypothesis generation rules from numerous known facts or relations extracted from the scientific literature. Each rule explicitly states the meaning of an association as a predicate and is able to produce an explanatory hypothesis in the form of "N_1 V N_2", where N and V denote a concept (noun phrase) and a predicate (verb phrase), respectively. Note that "N_1 V N_2" must not have been reported before to be considered as a hypothesis.

The second problem is the large number of generated hypotheses. Typically, most of the hypotheses are spurious and only a small fragment is worth further investigation. Because the latter is far outnumbered by the former and thus is difficult to find, it is crucial to prioritize or rank the hypotheses according to their plausibility. To this end, we first identify true associations among the automatically generated hypotheses and learn their characteristics by adopting a semi-supervised regression model. To build an effective model, we explore several types of features, including the reliability of the hypothesis generation rules, the semantic similarities between concepts, and specificity of concepts.

To demonstrate the validity of the proposed framework for hypothesis discovery, we carry out a series of experiments on the Medline database. The results show that the hypothesis generation rules acquired from the proposed framework successfully produce true hypotheses and that exploiting various features associated with the generated hypotheses are beneficial to rank those hypotheses.

2   Related Work

2.1   Hypothesis Discovery

Swanson has argued the potential use of a literature to discover new knowledge that has implicitly existed but been overlooked for years. His discovery framework is based on a syllogism. That is, two premises, "A causes B" and "B causes C," suggest a potential association, "A causes C," where A and C do not have a known, explicit relationship. Such an association can be seen as a hypothesis testable for verification to produce new knowledge, such as the aforementioned association between Raynaud's disease and fish oil. For this particular example, Swanson manually inspected two groups of articles, one concerning Raynaud's disease and the other concerning fish oil, and identified premises that "Raynaud's disease is characterized by high platelet affregability, high blood viscosity, and vasoconstriction" and that "dietary fish oil reduces blood lipids, platelet affregability, blood viscosity, and vascular reactivity," which together suggest a potential benefit of fish oil for Raynaud's patients. Based on the groundwork, Swanson himself and other researchers developed computer programs to aid hypothesis discovery. The following summarizes some of the seminal works and recent developments.

Weeber et al. [38] implemented a system, called DAD-system, taking advantage of a natural language processing tool. The key feature of their system is the incorporation of the Unified Medical Language System (UMLS') Metathesaurus[2] for knowledge representation and pruning. While the previous work focused on words or phrases appearing in Medline records for reasoning, DAD-system maps them to a set of concepts defined in the UMLS Metathesaurus using MetaMap [3]. An advantage of using MetaMap is that it can automatically collapse different wordforms (e.g., inflections) and synonyms to a single Metathesaurus concept. In addition, using *semantic types* (e.g., "Body location or region") under which each concept is categorized, irrelevant concepts can be excluded from further exploration if particular semantic types of interest are given. This filtering step can drastically reduce the number of potential associations, enabling more focused knowledge discovery. Pratt and Yetisgen-Yildiz's system, LitLinker [25], is similar to Weeber's, also using the UMLS Metathesaurus but adopted a technique from association rule mining [1] to find two associated concepts.

Srinivasan [30] developed another system, called Manjal, for hypothesis discovery. A primary difference of Manjal from the others is that it solely relies on MeSH[3] terms assigned to Medline records, disregarding all textual information. MeSH is a controlled vocabulary consisting of sets of terms (MeSH terms) used to manually indexing articles in life science. Manjal conducts a Medline search for a given concept and extracts MeSH terms from the retrieved articles. Then, according to predefined mapping, each of the extracted MeSH terms is associated with its corresponding UMLS semantic types. Similar to DAD-system, the subsequent processes can be restricted only to the concepts under particular semantic types of interest, so as to narrow the search space of potential paths. In addition, Manjal uses the semantic types for grouping resultant concepts in order to help users browse the system output. With Manjal, Srinivasan demonstrated that most of the hypotheses Swanson had found were successfully replicated.

More recently, some researchers [8,10] adopted the notion of *discovery patterns* [17], such as "drug x INHIBITS substance y, substance y CAUSES disease z", that link pharmaceutical substances to diseases they are known to treat. Unlike the co-occurrence-based approaches described above, the discovery pattern-based approach exploits predicates to enable more efficient and effective hypothesis discovery. For instance, Cohen et al. [10] used an NLP system, called SemRep [26], to extract semantic relations (e.g., "Acetylcholine STIMULATES

---

[2] http://www.nlm.nih.gov/research/umls/.

[3] http://www.nlm.nih.gov/mesh/.

Nitric Oxide"), where both noun phrases and predicates were normalized according to the UMLS Metathesaurus and the UMLS Semantic Network[4], respectively. The extracted relations were then encoded as vectors in hyperdimensional space to be used for deriving discovery patterns based on known treatment relations between pharmaceutical agents and diseases. Lastly, the discovered patterns were used to generate hypotheses. Their approach is similar to the present study in that both take advantage of the predicate-argument structure. The important differences, however, are (1) we do not require (and is not limited to) known treatment relations and that (2) we prioritize generated hypotheses based on their plausibility in a (semi-)supervised learning framework.

2.2   Textual Entailment

There is a related area of research in natural language processing (NLP), called *Recognizing Textual Entailment (RTE)*. RTE aims to identify if text fragment A implicates another text fragment B. For example, "SCO won a lawsuit against IBM" entails "SCO sued IBM" [36]. To recognize this kind of texual entailment, one needs a textual entailment rule, for example, "x win lawsuit against y" → "x sue y". A number of approaches have been proposed to automatically acquire such rules from a text corpus [5, 36].

Another work in RTE is to recognize a pair of clauses which have a causal relationship or a pair of contradictory clauses [12, 15]. An example of the former is "increase in crime" → "heighten anxiety", and that of the latter is "increase in crime" and "decrease in crime". By collecting these pairs, one could infer a causal relation which is the inverse of the former, that is, "decrease in crime" → "diminish anxiety" [15]. In addition, there is an attempt to acquire an inference rule to recognize textual entailment from two clauses [27]. For instance, two clauses, "*Food* is made from *Ingredient*" and "*Ingredient* contains *Chemical*", together suggest texual entailment: "*Food* contains *Chemical*". The idea is in essence the same as Swanson's syllogism. However, these works intend to generate known, yet implicit, relations or to generate paraphrases, whereas hypothesis discovery including our work aims at generating scientific hypotheses that are not known to be true. Both are basically done through inductive and deductive reasoning; the former generalizes from observed facts to acquire general rules and the latter uses the rules to infer known facts or hypotheses as logical conclusions. Thus, their approaches are sometimes similar, as in the case of Schoenmackers et al. [27] and ours, both of which exploit a syllogistic pattern between two relations. Their difference mostly lies not in their approaches but in their intended applications, which is reflected in the respective system designs and evaluation methodologies.

For example, Schoenmackers et al. [27] abstract concepts to semantically higher semantic classes (e.g., *Food* and *Ingredient*) and attempt to identify RTE rules that hold for those semantic classes. On the other hand, it is desirable for hypothesis discovery to generate a concrete hypothesis involving "specific"

---

[4] http://semanticnetwork.nlm.nih.gov/.

concepts, where the abstraction of concepts is not suitable. Instead, focusing on Swanson's syllogism, we acquire hypothesis generation rules and exhaustively generate hypotheses, which are then ranked according to their plausibility estimated from various attributes associated with the rules and resulting hypotheses. In addition, Schoenmackers et al. evaluated their inference rules based on known facts found in their knowledge base. In contrast, we evaluate generated hypotheses against true hypotheses found in the most recent subset of our knowledge base which is not used in rule acquisition or parameter tuning. More details shall be presented in the following sections.

3   Proposed Framework

3.1   Overview

The proposed framework is roughly divided into two parts. One is *hypothesis generation* and the other is *learning true hypotheses*. The former first derives hypothesis generation rules based on known facts or relations represented in predicate argument structure extracted from a corpus of texts and then generates potential hypotheses by applying the acquired rules to known relations. The latter uses true hypotheses as positive examples for learning a regression model and applies it to the generated hypotheses for prioritization according to their plausibility. Figure 1 illustrates the overview of the framework. The following sections describe each component depicted in the figure by roughly following the flow of the processes/data.

3.2   Deriving Hypothesis Generation Rules

Much previous work generates hypotheses simply based on co-occurrences of two terms or concepts. Although such approaches may produce valid hypotheses, they also produce even more spurious ones, making it more difficult to spot truly important hypotheses. This study takes into account the meaning of the relation between two concepts instead of their simple co-occurrence and only produces more reasonable hypotheses in consideration of the existing knowledge. In this study, each known relation extracted from the existing knowledge is expressed as a predicate-argument structure "$N_1 \, V \, N_2$", where V is a predicate and N_1 and N_2 are its subjective and objective arguments, respectively. Based on commonsense arguments, these relations are merged to identify a *chain of relations* described shortly to derive a hypothesis generation rule.

Knowledge Extraction. To extract known relations from the literature, this study relies on publicly available NLP tools, specifically, a chunker and a named entity (NE) recognizer. (The present study utilizes GENIA tagger [37] which provides both functionalities). Based on the former's output, a predicate-argument structure corresponding to "NP VP NP" is identified using a simple finite-state automaton. Figure 2 shows the automaton to be used in this work, where the

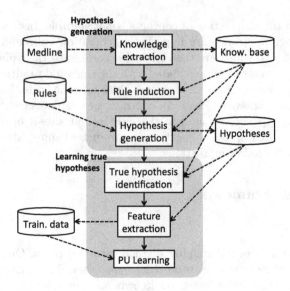

Fig. 1. Overview of the supervised hypothesis discovery framework. Solid and dotted lines show the flow of the processes and the flow of the data, respectively.

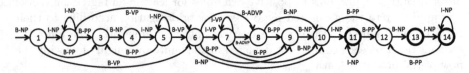

Fig. 2. Automaton to extract predicate-argument structures.

bold circles (11, 13, 14) indicate accepting states. The input of the automaton is a chunk tag (output of a chunker), such as B-NP and I-NP, indicating the begining of a noun phrase and inside of a noun phrase, respectively.

For example, from a sentence "the guideline is being reexamined currently by the NCRP committee", a relation ⟨the guideline, is being reexamined currently by, the NCRP committee⟩ is extracted. In addition, the following preprocessing is applied in this order to normalize the representation.

1. Transform all the terms to their base forms.
2. Remove all articles.
3. Replace negative adverbs (e.g., barely) with "not".
4. Remove all adverbs (except for "not").
5. Remove the relation itself if auxiliary verb is uncertain ("may" or "might").
6. Remove auxiliary verb.
7. Remove present/past/future tense.
8. Transform passive voice to active.

For the above example, the extracted relation is finally normalized to ⟨NCRP committee, reexamine, guideline⟩.

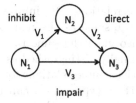

Fig. 3. A closed chain of relations leading to a hypothesis generation rule.

An NE recognizer identifies biomedical entities, including proteins and RNA. We retain only relations containing at least one such entity, which would help to generate biomedically meaningful hypotheses. To be precise, GENIA tagger employed in this work recognizes "Protein", "DNA", "RNA", "Cell line", and "Cell type" as NE types. Hereafter, the set of extracted relations is referred to as the knowledge base, denoted as **K**.

Rule Induction and Hypothesis Generation. From the knowledge base **K**, a hypothesis generation rule r is derived as a sequence of three predicates V_1, V_2, V_3. The basic idea is to identify a syllogistic pattern composed of three relations corresponding to two premises and one conclusion in Swanson's syllogism by merging the same arguments. For example, suppose that two relations were extracted from the literature: "N_1 inhibits N_2" and "N_2 directs N_3". The objective and subjective arguments of the former and latter, respectively, are the same (i.e., N_2) and thus form a chain of two relations by merging them: "N_1-inhibit-N_2-direct-N_3". Here, the previous work in hypothesis discovery may suggest that N_1 and N_3 have *some* implicit association without being able to specify the meaning of the association. In contrast, we take a further step to search for another relation involving N_1 and N_3, such as "N_1 impairs N_3" in the knowledge base **K**. This time, the subjective and objective arguments are the same as those of the first two relations, respectively. Further merging the arguments produces a triangular chain of relations as shown in Fig. 3.

These known relations collectively suggest that a general rule r below may hold through inductive generalization:

Rule r: If "x inhibits y" and "y directs z", then "x impairs z"[5],

where x, y, and z can be any noun phrases. Inductive generalization is a family of accounts of inductive inference and assumes that a single known instance confirms generalization [24].

Note that the rule only indicates a possible association that may be invalid. However, because the possible association follows a more reasonable logic than mere co-occurrence-based approaches, fewer spurious hypotheses are expected.

---

[5] In fact, three relations, "actinomycin D *inhibits* mRNA", "mRNA *directs* protein synthesis", and "actinomycin D *impairs* protein synthesis", were extracted from Medline, and this rule was acquired without manually coding any domain knowledge.

Note that if there exist three relations "N_a inhibit N_b", "N_b direct N_c", "N_a impair N_c" in **K** which are the same as the relations in Fig. 3 except for the arguments, these relations also yield the same rule (inhibit, direct, impair) as Fig. 3. However, they are internally distinguished as we keep track of their arguments involved in the respective source relations for later use.

These rules, denoted as $\mathbf{R} = \{r_1, r_2, \cdots\}$, can be easily identified by first finding two predicate-argument structures that share the same argument as the object and subject (i.e., N_2), and then finding another predicate-argument structure having the other two arguments (N_1 and N_3) as its subject and object. As mentioned, rule r also keeps the information on N_1, N_2, and N_3 for reasons described later. Once such rules **R** are exhaustively identified in the knowledge base **K**, they can be applied back to **K** to generate hypotheses, denoted as $\mathcal{H} = \{h_1, h_2, \cdots\}$ where h is a generated hypothesis. Let us stress that, in the hypothesis, the exact meaning of the association between two concepts is explicitly stated as a predicate (i.e., "impair" in the above example).

It should be mentioned that the above procedure would also generate hypotheses which already exist in the knowledge base **K**. To avoid it, generated hypotheses are first compared with existing relations in **K** and only those which do not exist are outputted as hypotheses.

3.3   Learning True Hypotheses

The number of hypotheses $|\mathcal{H}|$ to be generated from the rules **R** will be much smaller than co-occurrence-based approaches. Nonetheless, the number will be still high, where randomly investigating each hypothesis is time-consuming and costly. Therefore, it is crucial to prioritize or rank the generated hypotheses by considering their plausibility. To this end, we attempt to learn the characteristics of the "true" hypotheses using a supervised learning framework and predict a plausibility or confidence of each generated hypothesis h. Specifically, we take the following three steps: identification of true hypotheses, feature extraction, and PU learning, each described below.

Identification of True Hypotheses. For applying supervised learning, there need to be labeled examples, i.e., true and false hypotheses in this case. Such labeled examples are often manually created in many classification/regression tasks, such as spam filtering. However, it is not realistic to manually judge the validity of the generated hypotheses since it may require domain expertise, extensive reading, and even laboratory experiments. Instead, we take advantage of the biomedical literature more recent than those used for rule induction and hypothesis generation. In other words, if a generated hypothesis is found in the recent literature, the hypothesis can be seen as a validated, true hypothesis.

Specifically, we first split the knowledge base **K** into three subsets: $\mathbf{K}_1, \mathbf{K}_2, \mathbf{K}_3$ with $\mathbf{K}_1$ being the oldest set of knowledge and $\mathbf{K}_3$ the most recent. Then, $\mathbf{K}_1$ is used for inducing the rules, denoted as $\mathbf{R}_1$, and then for generating hypotheses, denoted as $\mathcal{H}_1$. Among $\mathcal{H}_1$, true hypotheses are identified using more recent knowledge K_2. Note that the remaining, newest knowledge K_3 will be held for evaluation as described later.

A potential problem of this approach is that the hypotheses not found in K_2 cannot be simply regarded as false hypotheses. This is because it is possible that they are actually true hypotheses but not yet appear in the literature. In other words, definite negative examples are difficult to identify. To cope with this issue, we see the non-true, inconclusive hypotheses as "noisy" data containing both true and false hypotheses and adopt a learning approach using positive and unlabeled examples, so called *PU learning*. Specifically, this study adopts an existing approach proposed by Lee and Liu [20], which considers PU learning as a problem of learning with noise by labeling all unlabeled examples as negative and learns a linear regression model.

Feature Extraction. To apply a supervised learning method, each hypothesis (instance) needs to be represented by a set of features that possibly reflect the characteristics of true/false hypotheses. There are two types of information that can be used to predict the plausibility of hypothesis h generated by a rule r. One is associated with r itself and the other is associated with r and h. In the following, the former is called *rule-dependent* features, and the latter *rule/hypothesis-dependent* features. Note that a rule is represented by a sequence of verbs (V_1, V_2, V_3) but is associated with the noun phrases (N_1, N_2, N_3) in the syllogistic pattern from which the rule is derived so as to obtain some of the features.

For the rule-dependent features, this work uses the ones summarized below.

- The number of syllogistic patterns that resulted in the same rule (V_1, V_2, V_3). The rationale is that if multiple patterns lead to the same rule, it is thought to be more reliable. Note that two rules, (V_a, V_b, V_c) and (V_d, V_e, V_f), are considered the same if and only if $V_a = V_d$, $V_b = V_e$, and $V_c = V_f$.
- Specificity of verbs. More specific verbs may lead to more specific, useful hypotheses. Following the intuition, two features below are extracted for each verb, V_1, V_2, and V_3, involved in a rule r (see Fig. 3).

  - Document frequency (DF) in Medline. The inverse of DF is often used as an indicator of the specificity of a word in information retrieval [29].
  - The number of synonyms. We assume that broader terms have more synonyms, and *vice versa*. The number is based on an English lexical database, WordNet [14].

- Specificity of nouns. The assumption is similar to the above. For each noun, N_1, N_2, and N_3, involved in a rule r, features below are extracted. Note that when the same r is derived from multiple syllogistic patterns, there are multiple sets of $\langle N_1, N_2, N_3 \rangle$. In that case, the generated hypothesis is replicated for each pattern so as to encode different feature values.

  - DF in Medline.
  - The number of synonyms.

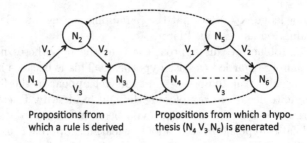

Propositions from   Propositions from which a hypo-
which a rule is derived   thesis (N₄ V₃ N₆) is generated

Fig. 4. Analogical resemblance between two chains of relations. A dashed line connects two concepts playing the same role in the rule represented by V_1, V_2, and V_3.

For the rule/hypothesis-dependent features, the followings are utilized.

- The number of rules that produced the same hypothesis h. The rationale is that if multiple rules produced the same hypothesis, it is thought to be more reliable.
- Specificity of nouns involved in the premises "N_4 V_1 N_5" and "N_5 V_2 N_6" that produced h (see the right-hand side of Fig. 4). For each noun (i.e., N_4, N_5, and N_6), its DF in Medline is used as the specificity.
- Applicability of rule r in generating hypothesis h. We assume that h is more plausible if r which generated h is more appropriate to the context (two premises) to be applied. We define this applicability of r as "analogical resemblance" between the syllogistic pattern from which r was derived and the one associated with h. The details are described in the next paragraphs.

Analogical Resemblance. Figure 4 illustrates the idea of analogical resemblance, where the left triangle is the syllogistic pattern from which a rule is derived and the right triangle is the two premises "N_4 V_1 N_5" and "N_5 V_2 N_6" from which a possible hypothesis "N_4 V_3 N_6" is inferred. A dashed line connects two concepts that play the same role in the syllogistic rule represented by a sequence of predicates V_1, V_2, and V_3. If the connected concepts are semantically more similar to each other, the rule is likely to be more applicable to the right triangle with concepts N_4, N_5, and N_6.

There is much work in estimating the semantic similarity of two concepts, such as corpus-based and lexicon-based approaches [22]. This study adopts a corpus-based approach for its wider coverage, specifically, *Normalized Google Distance* (NGD) [9]. NGD is an approximation of Normalized Information Distance (NID) and replaces the Kolmogorov complexity in the formulation with the number of Google hits as defined in

$$\text{NGD}(x,y) = \frac{\max\{\log f(x), \log f(y)\} - \log f(x,y)}{\log N - \min\{\log f(x), \log f(y)\}}, \tag{1}$$

where $f(x)$ is the number of hits by Google search with query x and N is the number of web pages indexed by Google. $\text{NGD}(x,y)$ ranges from 0 to ∞ and $\text{NGD}(x,y)=0$ means that they are identical.

Instead of Google, however, we use Medline, which would better reflect the domain knowledge and also ensures that $f(x)$ for any concept x will exist (non-zero) since the concepts in our knowledge base are all extracted from Medline in this study. Although the formulation is exactly the same, we call the distance used with Medline *Normalized Medline Distance* (NMD) to avoid unnecessary confusion. It should be mentioned that Lu and Wilbur [21] also used Medline for computing NGD. We compute an NMD value for each of the three pairs of concepts in Fig. 4, i.e., NMD(N_1,N_4), NMD(N_2,N_5), NMD(N_3,N_6), to represent the applicability of rule r to hypothesis h.

4 Evaluation

4.1 Experimental Procedure and Settings

To demonstrate the validity of our proposed framework, we performed evaluative experiments. As the existing, public knowledge, we used a subset of Medline, specifically, the 2004 Genomics track data set [16]. The data set is a subset of Medline from 1994 to 2003 and is composed of 4,591,008 records. From the data set, known relations were extracted in a predicate-argument structure from the titles and abstracts of the Medline records using the GENIA tagger [37]. After applying the normalizing processes described in the *Proposed Framework* section, 17,904,002 relations were acquired, which formed our knowledge base **K**. Then, **K** was split into three subsets $\mathbf{K}_1, \mathbf{K}_2, \mathbf{K}_3$ of around the same size. The oldest knowledge, $\mathbf{K}_1$, was used for rule derivation and hypothesis generation. The number of rules was 12,180 and the number of generated hypotheses was 346,424 including duplicates.

The generated hypotheses were then compared with the knowledge base $\mathbf{K}_2$ and subsequently $\mathbf{K}_3$. If a hypothesis was found in $\mathbf{K}_2$, it was considered as a positive example and was added to the training data. If it was not found in $\mathbf{K}_2$ but found in $\mathbf{K}_3$, it was added to the test data as a positive example. The hypotheses not found in $\mathbf{K}_2$ nor $\mathbf{K}_3$ were unlabeled and were added to either the training or test data at random. This process ensures that training and test data do not have the same instances.

The training data were used for PU learning and the test data were used for evaluating the performance of our proposed framework for hypothesis generation and ranking. The training data contain 226 positive and 169,060 unlabeled examples, and the test data contain 88 positive and 169,059 unlabeled examples. The unlabeled examples in the test data are regarded as negatives in evaluation, although they may not be truly negatives as they may be verified in future. We will come back to this issue in the next section when discussing the evaluation criteria.

For PU learning, we adopted the algorithm proposed by Lee and Liu [20]. PU learning needs positive examples and unlabeled examples as training data. Lee and Liu's algorithm considers PU learning as a problem of learning with noise and learns a logistic regression model by optimizing a cost function approximating the actual target function (expected weighted error) by gradient descent.

To be precise, the following update function is iterative applied to estimate weight w for each feature:

$$w_{j,t} = (1 - c)w_{j,t-1} + \eta(\Delta_{j,t} + \gamma\Delta_{j,t-1}) \tag{2}$$

where j is the index of features, t denotes an epoch, c is a decay parameter, η is a learning rate, $\Delta_{j,t}$ is the j-th component of the negative gradient of the cost function at epoch t, and γ is a momentum parameter. Note that this model uses a cost function considering the number of positive and unlabelled examples so that it can make robust predictions even for unbalanced data. The parameters, t, c, η, and γ were set to 500, 0.05, 0.00001, and 0.99, respectively, by consulting the original paper by Lee and Liu [20].

4.2   Empirical Results

Generated Hypotheses. Among the generated hypotheses, Tables 1 and 2 show 20 examples of false and true hypotheses, respectively (i.e., the former were not found in K_3 and the latter were), where predicates are italicized.

Table 1 includes hypotheses which do not make sense and are clearly invalid, such as "datum *be escape from* expression of cox 2". On the other hand, Table 2 provides hypotheses that are re-discovery of true relations found in the test data (knowledge base K_3). Comparing the true and false hypotheses, the former contains relatively shorter noun phrases than the latter. This reflects the fact that hypotheses are checked against the knowledge base by surface forms (exact match) and shorter phrases simply have more chance to be matched. There may be true hypotheses which have different surface forms from the relations in our knowledge base, and ideally their identity should be recognized. To this end, SemRep which extracts relations expressed in UMLS Metathesaurus/Semantic Network concepts, or the techniques developed for paraphrasing and RTE would be beneficial.

It should be stressed that hypotheses generated by the previous works using co-occurrences of terms or concepts could not indicate the nature of the relations. In other words, only a pair of terms/concepts (e.g., "LPS (lipopolysaccharide)" and "DC (dendritic cells)") are suggested to be potentially related. One needs to interpret the relation between the two by own expertise or researching abundant literatures. In contrast, our proposed framework suggests concrete hypotheses with specific relations as predicates (e.g., "LPS *stimulate* DC") and thus there is no need for interpretation.

Performance of Hypothesis Generation and Ranking. The generated hypotheses in the test data were ranked by the output of the regression model learned from the training data. The performance of the ranking was evaluated by a receiver operating characteristic (ROC) curve and the area under it (AUC). An ROC curve is plotted with x and y axes being false positive and true positive rates, respectively, and is often used for evaluating the performance of classifiers.

Table 1. Examples of generated false hypotheses.

cross recognition of lps *play* predominant role in salmonella pathogenesis
ppt injection *be escape from* relaxation in control
second embolization *regulate* sclerosis of adjacent laryngeal cartilage
dysplastic epithelium *be inject into* apical portion of luminal cell
occlusion by intraluminal filament technique *trigger* proliferation of thyroid cell
homozygote *play* important role in development of common disease among northern indigenous people
introduction of neomycin resistance gene cartridge in cod region *inhibit* exchange of anion
mutate protein *play* important role as co factor in disease transmission
gamma proteobacteria symbiont *be partition at* 3 to 5 degree
modification to receptor *stabilize* volume
expression *induce* tnf alpha as result of cellular oxidative stress
moderate level of noise exposure *antagonize* abdominal fat measurement
testicular tissue *be amplify from* ws
essential definition of term *take* place in outpatient clinic of 3 veterans administration hospital
datum *be escape from* expression of cox 2
complete resection with mediastinal lymphadenectomy *regulate* favourable view on vocational training
mean of co2 laser vaporization *regulate* successful clinical
surgical decompression *regulate* three identical antigen
wide variety of carcinogen *be mediate at* usable frequency in cell
biliary cirrhosis and tsunoda type iii and iv *induce* analysis

There are other commonly used evaluation criteria, including accuracy and F-measure. However, accuracy is not suitable for this experiment as the number of positive and negative examples is heavily unbalanced. F-measure is not suitable either because negative examples in the test data may not be actually negatives. An ROC curve is more appropriate in this setting. It basically indicates the degree to which the distributions of positives and negatives are separated, and even if some negatives are actually positives, the influence on the resulting ROC curve is limited if the number of such not-yet-known positives is much smaller than that of the negatives.

As a baseline, we used the ranking approach by Hristovski et al. [18], which ranks the generated hypotheses based on the confidence values used in association rule mining. Specifically, the confidence of y with respect to x, denoted as $C_{x \to y}$, is defined as the ratio of the number of documents containing x to the number of documents containing both x and y. Similarly, the confidence of z with respect to y, i.e., $C_{y \to z}$, is computed. Hristovski et al. [18] used their

Table 2. Examples of generated true hypotheses.

t lymphocyte *produce* interleukin 2
actin *activate* atpase activity
lymphocyte *produce* cytokine
seb *induce* lethal shock
osteoclast *produce* reactive oxygen intermediate
vanadate *induce* contraction
radiation *induce* necrosis
hypertension *induce* vascular disease
glucose *induce* phosphorylation of insulin receptor
hydrogen peroxide *induce* necrosis
macrophage *produce* cytokine
fat *suppress* mri
radiotherapy *induce* mucositis
protease *activate* g protein
lps *stimulate* pbmc
endothelium *produce* oxygen
ethanol *induce* locomotor activity
scf *induce* proliferation
tbid *induce* cytochrome c release
lps *stimulate* dc

product $C_{x \to y} \times C_{y \to z}$ to rank hypotheses. Figure 5 shows the ROC curves for the baseline (Confidence), our proposed framework (PU learning), where the diagonal corresponds to random guess (Random).

The baseline (Confidence) and our framework (PU learning) achieved the AUC of 0.769 and 0.860, respectively. While both outperformed the random guess (Random) whose AUC is 0.5, our framework showed greater performance. This is presumably due to the fact that our framework leverages not only frequencies of documents used for computing confidence but also other features including the specificity of predicates/nouns and semantic relatedness between chains of relations and applies PU learning to estimate model parameters (feature weights) as discussed below.

We then looked at which features contributed to the performance based on the regression coefficients (feature weights) as summarized in Table 3, where features are sorted in descending order of the absolute weight values within the rule-dependent and rule/hypothesis-dependent groups.

Let us first examine the rule-dependent features. Among them, DF of V_3 was found to have the greatest predictive power with the highest weight of 246.33, followed by DF of N_2 and DF of V_1 and other DF values. The higher effect of V_3's makes sense as it appears as the predicate of the generated hypotheses.

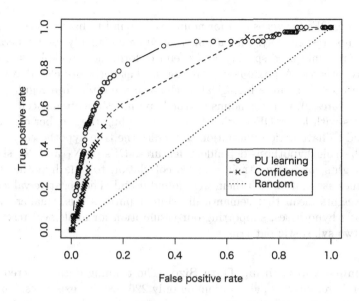

Fig. 5. Performance of hypothesis discovery.

Table 3. Comparison of the features in terms of the regression coefficients in the learned model.

	Feature	Regression coefficient
Rule-dependent	DF of V_3	246.33
	DF of N_2	−176.02
	DF of V_1	150.17
	DF of N_3	−118.27
	DF of N_1	−93.26
	DF of V_2	7.41
	# of patterns for the same rule	4.92
	# of synonyms of V_3	3.70
	# of synonyms of V_1	−2.64
	# of synonyms of N_2	−0.72
	# of synonyms of N_3	−0.17
	# of synonyms of V_2	0.17
	# of synonyms of N_1	−0.05
Rule/hypothesis-dependent	DF of N_4	43.40
	DF of N_6	39.64
	NMD b/w N_1 and N_4	−28.20
	NMD b/w N_3 and N_6	−21.73
	DF of N_5	−14.07
	NMD b/w N_2 and N_5	−9.03
	# of same hypothesis	−0.47

Interestingly, DFs for verbs and for nouns were found to have positive and negative weights, respectively. This means that more commonly used general verbs and less commonly used specific nouns tend to form more reliable rules leading to true hypotheses. A possible explanation is that verbs are closed vocabulary and those used to express biological mechanism are often in a regular pattern (e.g., x activates y), whereas nouns in true hypotheses are often specific biomedical entities with lower DFs. Other features including the number of synonyms were found to have little information to predict the true hypotheses.

For the rule/hypothesis-dependent features, DFs of N_4 and N_6 show the higher weights, indicating their positive correlation to true hypotheses. Also, NMD values associated with them were found useful. The negative values of the NMD's weights mean that semantically similar phrases (i.e., smaller distance) lead to true hypotheses, supporting our assumption for analogical resemblance between two syllogistic patterns.

Performance vs. Training Data Size. The training data collected for the evaluation is relatively small, containing only 226 positive examples. Therefore, it is important to investigate the effect of the training data size in learning a regression model to see if more training data would help to increase the performance. For this purpose, we randomly sampled $n\%$ of the training data and used them for learning a regression model. For each n, the same process was repeated for 10 times to compute an average AUC for the particular n. Figure 6 shows the transition of the average AUC with different values of n, where the error bars indicate ± 1 standard deviation.

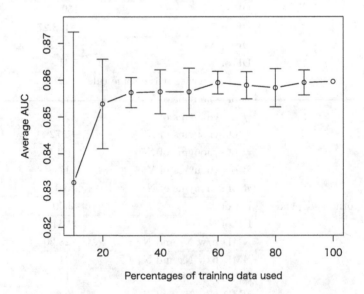

Fig. 6. Average AUC for different amount of training data.

The result shows that more training data gradually improve the performance but the difference is subtle. From this experiment, we conclude that more training data would not lead to performance boost with the current model and features. We plan to explore alternative PU learning models [13, 39] and richer features.

4.3 Discussion

While the proposed rule-based framework was shown effective in hypothesis discovery and ranking, the hypothesis generation rules based on simple predicate-argument structure (e.g., "N_1 V N_2") have inherent defects in both accuracy and coverage. Especially, our framework relies on surface forms of entities, not recognizing their spelling variations, aliases, or classes.

Beside using an automatically created or hand-crafted thesaurus (e.g., UMLS Metathesaurus) mentioned in Sect. 4.2, generative models would be useful to this end. For instance, latent Dirichlet allocation (LDA) [7] allows us to analyze underlying latent topics of term occurrences. Using such models, and by feeding the model with a large number of documents, one could identify related terms belonging to the same topics or sharing a similar topic mixture.

Deep learning [4], which has been attracting much attention in many research areas, would be beneficial in hypothesis discovery as well. Mikolov et al. [23] used a neural network language model to represent words in a vector space, which captures syntactic and semantic regularities of words. This model is particularly interesting in that it enables vector-oriented reasoning based on the offset between word vectors, such as "king − man + woman = queen". Beyond representing single words in a vector space, Socher et al. [28] developed recursive deep models to capture compositional effects of a longer word sequence, i.e., a phrase and a sentence. The model can perform sentiment analysis for an input sentence at word, phrase, and the whole sentence level. This model could be used to represent predicate-argument relations in a vector space, possibly allowing vector-oriented hypothesis discovery. Incorporating generative models and/or deep learning into our framework may advance the research in hypothesis discovery.

5 Conclusion and Future Work

This paper focused on a triangular chain of relations, called syllogistic patterns, and proposed a novel approach to hypothesis discovery. The key intuition is that a generalized rule can be induced from such patterns and can then be applied back to the existing knowledge to generate hypotheses. To validate the idea, we implemented the proposed framework and exhaustively identified such hypothesis generation rules in a subset of Medline database and generated hypotheses based on the acquired rules. Among them, true hypotheses were automatically identified based on more recent literature so as to construct training/test data for PU learning. We examined various features associated with specificity, analogical resemblance, and others to represent generated hypotheses. Our evaluation demonstrated that the proposed framework was effective in discovering true hypotheses and that some of those features were characteristic to true hypotheses.

Although the results are promising, the present work has several limitations. The literature used for our experiment was limited in the amount and coverage. Thus, the identified true hypotheses may be actually old knowledge that simply did not appear in our data. Also, basic biomedical knowledge may not appear in Medline and thus not in our knowledge base, either. Furthermore, the current knowledge extraction scheme acquires simple SVO triplets, removing all nuance and all context, which results in rather vague relations, such as "ethanol induce locomotor activity" in Table 2, which can be both true and false depending on dose, species, etc. To better understand the issue, the generated hypotheses need to be qualitatively analyzed by biomedical experts. Another issue is concerned with hypothesis ranking. The generated hypotheses were ranked by plausibility, not interestingness. Therefore, the highly ranked hypotheses are not necessarily interesting or surprising for experts. These issues should be tackled in the future work.

Acknowledgements. This work is partially supported by JSPS KAKENHI Grant Numbers 25330363 and MEXT, Japan.

References

1. Agrawal, R., Mannila, H., Srikant, R., Toivonen, H., Verkamo, A., et al.: Fast discovery of association rules. Adv. Knowl. Discov. Data Min. **12**, 307–328 (1996)
2. Ananiadou, S., Kell, D.B., Tsujii, J.: Text mining and its potential applications in systems biology. Trends Biotechnol. **24**(12), 571–579 (2006)
3. Aronson, A.R.: Effective mapping of biomedical text to the UMLS metathesaurus: the metamap program. In: Proceedings of American Medical Informatics 2001 Annual Symposium, pp. 17–21 (2001)
4. Bengio, Y.: Learning deep architectures for AI. Found. Trends Mach. Learn. **2**(1), 1–127 (2009)
5. Berant, J., Dagan, I., Adler, M., Goldberger, J.: Efficient tree-based approximation for entailment graph learning. In: Proceedings of the 50th Annual Meeting of the Association for Computational Linguistics, pp. 117–125 (2012)
6. Björne, J., Ginter, F., Pyysalo, S., Tsujii, J., Salakoski, T.: Complex event extraction at PubMed scale. Bioinformatics **26**(12), i382–i390 (2010)
7. Blei, D.M., Ng, A.Y., Jordan, M.I.: Latent dirichlet allocation. J. Mach. Learn. Res. **3**, 993–1022 (2003)
8. Cameron, D., Bodenreider, O., Yalamanchili, H., Danh, T., Vallabhaneni, S., Thirunarayan, K., Sheth, A.P., Rindflesch, T.C.: A graph-based recovery and decomposition of Swanson's hypothesis using semantic predications. J. Biomed. Inf. **46**(2), 238–251 (2013)
9. Cilibrasi, R.L., Vitanyi, P.M.B.: The Google similarity distance. IEEE Trans. Knowl. Data Eng. **19**, 370–383 (2007)
10. Cohen, T., Widdows, D., Schvaneveldt, R.W., Davies, P., Rindflesch, T.C.: Discovering discovery patterns with predication-based semantic indexing. J. Biomed. Inf. **45**(6), 1049–1065 (2012)

11. Digiacomo, R.A., Kremer, J.M., Shah, D.M.: Fish-oil dietary supplementation in patients with Raynaud's phenomenon: a double-blind, controlled, prospective study. Am. J. Med. **86**(2), 158–164 (1989)

12. Do, Q.X., Chan, Y.S., Roth, D.: Minimally supervised event causality identification. In: Proceedings of the Conference on Empirical Methods in Natural Language Processing, pp. 294–303 (2011)

13. Elkan, C., Noto, K.: Learning classifiers from only positive and unlabeled data. In: Proceedings of the 14th ACM SIGKDD International Conference on Knowledge Discovery and Data Mining, pp. 213–220 (2008)

14. Fellbaum, C.D.: WordNet: an electronic lexical database. MIT Press, Cambridge (1998)

15. Hashimoto, C., Torisawa, K., De Saeger, S., Oh, J.H., Kazama, J.: Excitatory or inhibitory: a new semantic orientation extracts contradiction and causality from the Web. In: Proceedings of the 2012 Joint Conference on EMNLP/CoNLL, pp. 619–630 (2012)

16. Hersh, W., Bhuptiraju, R.T., Ross, L., Cohen, A.M., Kraemer, D.F.: TREC 2004 genomics track overview. In: Proceedings of the 13th Text REtrieval Conference (TREC) (2004)

17. Hristovski, D., Friedman, C., Rindflesch, T.C., Peterlin, B.: Exploiting semantic relations for literature-based discovery. In: Proceedings of American Medical Informatics 2006 Annual Symposium, pp. 349–353 (2006)

18. Hristovski, D., Peterlin, B., Mitchell, J.A., Humphreyb, S.M.: Using literature-based discovery to identify disease candidate genes. Int. J. Med. Inf. **74**, 289–298 (2005)

19. Kostoff, R.N., Block, J.A., Solka, J.L., Briggs, M.B., Rushenberg, R.L., Stump, J.A., Johnson, D., Lyons, T.J., Wyatt, J.R.: Literature-related discovery. Ann. Rev. Inf. Sci. Technol. **43**(1), 1–71 (2009)

20. Lee, W.S., Liu, B.: Learning with positive and unlabeled examples using weighted logistic regression. In: Proceedings of the 20th International Conference on Machine Learning (2003)

21. Lu, Z., Wilbur, W.J.: Improving accuracy for identifying related PubMed queries by an integrated approach. J. Biomed. Inf. **42**(5), 831–838 (2009)

22. Mihalcea, R., Corley, C., Strapparava, C.: Corpus-based and knowledge-based measures of text semantic similarity. In: Proceedings of the 21st National Conference on Artificial Intelligence, pp. 775–780 (2006)

23. Mikolov, T., Yih, W., Zweig, G.: Linguistic regularities in continuous space word representations. In: Proceedings of the 2013 Conference of the North American Chapter of the Association for Computational Linguistics: Human Language Technologies (NAACL-HLT-2013), pp. 746–751 (2013)

24. Norton, J.D.: A Little Survey of Induction. In: Achinstein, P. (ed.) Scientific Evidence: Philosophical Theories and Applications, pp. 9–34. Johns Hopkins University Press, Baltimore (2003)

25. Pratt, W., Yetisgen-Yildiz, M.: Litlinker: capturing connections across the biomedical literature. In: Proceedings of the 2nd international conference on Knowledge capture, pp. 105–112 (2003)

26. Rindflesh, T.C., Fiszman, M.: The interaction of domain knowledge and linguistic structure in natural language processing: interpreting hypernymic propositions in biomedical text. J. Biomed. Inf. **36**(6), 462–477 (2003)

27. Schoenmackers, S., Etzioni, O., Weld, D.S., Davis, J.: Learning first-order Horn clauses from web text. In: Proceedings of the 2010 Conference on Empirical Methods in Natural Language Processing, pp. 1088–1098 (2010)

28. Socher, R., Perelygin, A., Wu, J., Chuang, J., Manning, C., Ng, A., Potts, C.: Recursive deep models for semantic compositionality over a sentiment treebank. In: Proceedings of the 2013 Conference on Empirical Methods in Natural Language Processing, pp. 1631–1642 (2013)

29. Jones, K.S.: Statistical interpretation of term specificity and its application in retrieval. J. Documentation **28**(1), 11–20 (1972)

30. Srinivasan, P.: Text mining: generating hypotheses from Medline. J. Am. Soc. Inf. Sci. Technol. **55**(5), 396–413 (2004)

31. Swanson, D.R.: Fish oil, Raynaud's syndrome, and undiscovered public knowledge. Perspect. Biol. Med. **30**(1), 7–18 (1986)

32. Swanson, D.R.: Two medical literatures that are logically but not bibliographically connected. J. Am. Soc. Inf. Sci. **38**(4), 228–233 (1987)

33. Swanson, D.R.: Migraine and magnesium: eleven neglected connections. Perspect. Biol. Med. **31**(4), 526–557 (1988)

34. Swanson, D.R.: Somatomedin C and arginine: implicit connections between mutually isolated literatures. Perspect. Biol. Med. **33**(2), 157–179 (1990)

35. Swanson, D.R., Smalheiser, N.R., Torvik, V.I.: Ranking indirect connections in literature-based discovery: the role of medical subject headings. J. Am. Soc. Inf. Sci. Technol. **57**(11), 1427–1439 (2006)

36. Szpektor, I., Dagan, I.: Learning entailment rules for unary templates. In: Proceedings of the 22nd International Conference on Computational Linguistics, pp. 849–856 (2008)

37. Tsuruoka, Y., Tsujii, J.: Bidirectional inference with the easiest-first strategy for tagging sequence data. In: Proceedings of HLT/EMNLP 2005, pp. 467–474 (2005)

38. Weeber, M., Klein, H., Jong-van den Berg, L.T.W., Vos, R.: Using concepts in literature-based discovery: simulating Swanson's Raynaud-fish oil and migraine-magnesium discoveries. J. Am. Soc. Inf. Sci. Technol. **52**(7), 548–557 (2001)

39. Xiao, Y., Liu, B., Yin, J., Cao, L., Zhang, C., Hao, Z.: Similarity-based approach for positive and unlabelled learning. In: Proceedings of the 22nd International Joint Conference on Artificial Intelligence, pp. 1577–1582 (2011)

An Analysis of Variance-Based Methods for Data Aggregation in Periodic Sensor Networks

Hassan Harb[1,3]($\boxtimes$), Abdallah Makhoul[1], David Laiymani[1],
Oussama Bazzi[2], and Ali Jaber[3]

[1] FEMTO-ST Laboratory, DISC Department, University of Franche-Comté,
Belfort, France
{hassan.moustafa_harb,abdallah.makhoul,david.laiymani}@univ-fcomte.fr
[2] Department of Physics and Electronics, Lebanese University, Beirut, Lebanon
{obazzi,ali.jaber}@ul.edu.lb
[3] Department of Computer Science, Lebanese University, Beirut, Lebanon

Abstract. Given the vast area to be covered and the random deployment of the sensors, wireless sensor networks (WSNs) require scalable architecture and management strategies. In addition, sensors are usually powered by small batteries which are not always practical to recharge or replace. Hence, designing an efficient architecture and data management strategy for the sensor network are important to extend its lifetime. In this paper, we propose energy efficient two-level data aggregation technique based on clustering architecture with which data is sent periodically from nodes to their appropriate Cluster-Heads (CHs). The first level of data aggregation is applied at the node itself to eliminate redundancy from the collected raw data while the CH searches, at the second level, nodes that generate redundant data sets based on the variance study with three different Anova tests. Our proposed approach is validated via experiments on real sensor data and comparison with other existing data aggregation techniques.

Keywords: Periodic Sensor Networks (PSNs) · Data aggregation · Clustering architecture · Identical nodes behaviour · One way Anova model

1 Introduction

Wireless Sensor Networks (WSNs) have become one of the innovative technologies that are widely used nowadays. One of the advantages of these networks is their ability to operate unattended in harsh environments in which contemporary human-in-the-loop monitoring schemes are risky, inefficient and sometimes infeasible [1]. With the capabilities of pervasive surveillance, WSN have attracted significant attention in many applications, such as habitat monitoring [2], environment monitoring [3,4] and military surveillance [5,6]. In such networks, sensors are expected to be remotely deployed, e.g. via helicopter or clustered bombs,

© Springer-Verlag Berlin Heidelberg 2015
A. Hameurlain et al. (Eds.): TLDKS XXII, LNCS 9430, pp. 165–183, 2015.
DOI: 10.1007/978-3-662-48567-5_6

in a wide geographical area to monitor the changes in the environment and send back the collected data to a specific node called the "sink". Nevertheless, sensors in such environment are energy-constrained and their batteries cannot be replaced. Therefore, it is very important to limit the energy consumption of sensors in order to extend the network's lifetime as long as possible.

Due to a random and dense deployment, nodes may have overlapping sensing ranges, such that events can be detected by multiple sensor nodes providing a redundancy in sensed data. Moreover, since data transmission is more demanding than computational operations in terms of energy consumption, the volume of data transmitted must be minimized. This leads to the requirement of better data aggregation and data mining techniques. To that effect, data aggregation has been proved as an effective method to achieve power efficiency by reducing data redundancy and minimizing bandwidth usage [7] while data mining deals with extracting knowledge from large continuous arriving data from WSNs [8].

On the other side, clustering is considered as an efficient topology control method in WSN, which can increase network scalability and lifetime [9]. With clustering, data collected by sensor nodes are processed at intermediate nodes, called Cluster-Heads (CHs), in order to eliminate redundancy and send only the useful information to the sink (Fig. 1). In this paper, we use the periodic data collection approach, in which each sensor node sends periodically (at each period p) its data to the appropriate CH. We propose an energy efficient two-level data aggregation technique which applies at each cluster separately. The first level is applied at the sensor node itself in order to eliminate redundancy from data collected by the sensor at each period p before sending them to their proper CH. Then, when the CH receives data from all its members (nodes) we propose to use the one way Anova model with three different tests (Fisher, Tukey and Bartlett) to detect nodes with identical behaviour which generate redundant data logs or sets. The aim is to reduce data redundancy generated by neighboring nodes based on the variance study in order to eliminate redundancy before sending final data to the sink.

The rest of this paper is organized as follows; Sect. 2 presents related work on data aggregation in the sensor networks. Section 3 describes the first phase of our technique which we called member node aggregation. In Sect. 4, we present the second level, called CH aggregation, which is based on one way Anova model. Experimental results are exposed in Sect. 5. Finally, we conclude our paper and we provide our directions for future work in Sect. 6.

2   Related Work

In WSN, many data aggregation studies have been made based on clustering schemes, such as DDCD [10] and DUCA [11]. The main objective of these works is balancing and reducing energy consumption over the whole network. In each cluster, the sensors communicate data to their CH that aggregates data and thus reduces the size of data to be transmitted to the sink. Recently, The authors in [12,13] present a comprehensive overview about different data aggregation techniques and clustering routing protocols proposed in the literature for WSNs.

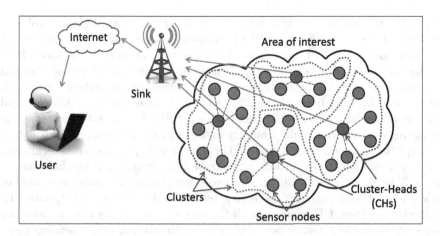

Fig. 1. Wireless sensor network based on two-tier single-hop clustering architecture.

The authors in [14] propose a Distributed K-mean Clustering (DKC) method for WSN. On the basis of DKC, the authors build a network data aggregation processing mechanism based on adaptive weighted allocation of WSN. DKC algorithm is mainly used to process the testing data of bottom nodes in order to reduce the data redundancy. In [15], the authors propose a data aggregation based clustering scheme for underwater wireless sensor networks (UWSNs) which involves four phases. The goals of these phases are to reduce the energy consumed in the overall network, increasing the throughput, and minimizing data redundancy. The authors in [16] propose a M-EECDA (Multihop Energy Efficient Clustering & Data Aggregation Protocol for Heterogeneous WSN). The protocol combines the idea of multihop communications and clustering for achieving the best performance in terms of network life and energy consumption. M-EECDA introduces a sleep state and three tier architecture for some cluster heads to save energy in the network.

Some other works in data aggregation are not based on clustering scheme: The authors in [17] propose a structure-free and energy-balanced data aggregation protocol, SFEB. SFEB features both efficient data gathering and balanced energy consumption, which result from its two-phase aggregation process and the dynamic aggregator selection mechanism. In [18], the authors propose an automatic auto regressive-integrated moving average modeling-based data aggregation scheme in WSNs. The main idea behind this scheme is to decrease the number of transmitted data values between sensor nodes and aggregators by using time series prediction model. In [19], the authors study the problem of building maximum lifetime shortest path aggregation trees in WSNs. When the shortest path trees are built, the authors transformed the problem into a load balancing scheme at each level of the fat tree and solved it by a centralized approach in polynomial time. The authors in [20] propose a data aggregation with multiple sinks in an Information-Centric Wireless Sensor Network with an ID-based information-centric network, in order to reduce the energy-transmission cost.

In [21], the authors study a new area within filtering aggregation problem, the Prefix-Frequency Filtering (PFF) technique. Further to a local processing at sensor node level, PFF uses Jaccard similarity function at aggregator's level to identify similarities between near sensor nodes and integrate their sensed data into one record. Aiming to decrease data latency, the authors in [22,23] propose two optimizations of the PFF technique based on suffix filtering and k-means algorithm. Among all optimizations, PFF stays a hard technique for the aggregator in terms of data latency and energy consumption. In this paper, we adapt the same scenario as proposed in [21] while we propose a new technique. In the new technique, we propose a two-level data aggregation, the first one, at the node level, which we call member node aggregation in which each member node sends, at each period p, its aggregated set of data to the appropriate CH. At the second level, CH aggregates all the sets of data coming from its member nodes based on the variance between their measurements, before sending them to the sink.

3   First Level: Member Node Aggregation

In periodic sensor networks (PSNs), each sensor node i takes a new measurement y_{is} at each time slot s. Then node i forms a new vector of captured measurements $M_i = [y_{i_1}, y_{i_2}, \ldots y_{i_T}]$ at each period p, where T is the total number of measures taken at the period p, and sends it to the appropriate CH [21]. Figure 2 shows an example of PSN where each sensor node takes one data measurement each ten minutes, e.g. $s = 10$ minutes, and send its set of collected data which contains six measures, e.g. $T = 6$, to the CH at the end of each hour.

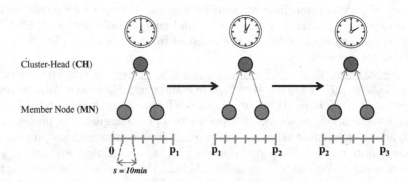

Fig. 2. Illustrative example of periodic sensor network (PSN).

Consequently, one of the important design considerations associated with the periodic sampling data model is that the dynamics of the monitored conditions can slow down or speed up [24]. Thus, it is likely that a sensor node takes the same (or very similar) measurements several times, especially when s is too short, which make the sensor node forwards more redundant data to the CH during each period. In this phase of aggregation, which called member node aggregation, we allow each sensor node to identify and remove duplicate data measurements among data collected in each period in order to reduce the size of the set M_i before sending it to the CH. In order to identify the similarity between two measures, we provide the two following definitions:

Definition 1 (*Similar* **function**). *We define the Similar function between two measurements as:*

$$Similar(y_i, y_j) = \begin{cases} 1 & \text{if} \quad \|y_i, y_j\| \leq \delta, \\ 0 & \text{otherwise.} \end{cases}$$

where δ is a threshold determined by the application. Furthermore, two measures are similar if and only if their *Similar* function is equal to 1.

Definition 2 (**Measure's weight,** $wgt(y_i)$). *The weight of a measurement y_i is defined as the frequency of the same or similar (according to the Similar function) measurements in the same set.*

For each new sensed measurement (at each slot s), a sensor node i searches for the similar measure already captured in the same period p. If a similar measurement is found, the sensor deletes the new measure while incrementing the weight of the existing measure by one, else, the sensor adds the new measure to the set and initializes its weight to 1. For more details about this algorithm see Algorithm 1 in [21].

Based on the above definitions, we provide two other definitions:

Definition 3 (**Cardinality of the set** M_i, $|M_i|$). *The cardinality of the set M_i is equal to the number of elements in M_i.*

Definition 4 (**Weighted Cardinality of the set** M_i, $Card_w(M_i)$). *The weighted cardinality of the set M_i is equal to the sum of all measures' weights in M_i as follow: $Card_w(M_i) = \sum_{k=1}^{|M_i|} wgt(m_k)$, where $m_k \in M_i$.*

In this paper, we consider that all sensor nodes operate at the same sampling rate, and every node captures T measures in each period p. Thus we can deduce that for every received set M_i from node i we have: $Card_w(M_i) = T$.

At the end of each period p, each member node i will possess a set of reduced measures associated to their corresponding weight. The second step is to send it to the appropriate CH which in its turn aggregates the data sets coming from different member nodes.

4   Second Level: CH Aggregation

At this level of aggregation, each CH receives all the sets of measurements with their weights sent from its member nodes, at the end of each period. The idea is to identify all pairs of member nodes that generate redundant sets in order to eliminate duplication before sending them to the sink. Therefore, one way

Anova model is an effective technique that can determine duplicated sets based on the variance between their measures. The Anova produces an F-statistic, the ratio of the variance calculated based on the measurements in the sets. F can be calculated in different manners depending on the statistic tests proposed in the Anova model. The sets are considered duplicated if the calculated F is less than the critical value of the F-distribution (or F_s) for some desired false-rejection probability (risk α). In [24], tha authors used one way Anova model and Fisher test in PSN at the level of node member to adapt its sampling rate. In this paper, we use the one way Anova model at the CH level, while comparing three different tests (Fisher, Tukey and Bartlett) in order to identify identical nodes behaviour.

4.1   One-Way ANalysis Of VAriance: ANOVA

In this part, we present a statistical model to study the variance between measurements in the data sets in order to find all pairs of member node that generate redundant data. Therefore, one-way Anova is used to find out if the means of data sets are significantly different or if they are relatively the same. In PSN, we assume that each sensor node takes T measures of temperature or humidity within a period p.

When receiving data sets coming from its member nodes at each period, CH computes the variation between every pair of sets. Therefore, it uses the one way Anova to test whether or not the means of every pair are equal. In case that a pair of sets notices low differences variance, CH considers that the two member nodes generate redundant data. After identifying all pairs of redundant sets, CH uses selecting sets algorithm proposed in the later subsection to select final sets to be sent to the sink, while conserving the integrity of information.

We suppose that measures generated by each member node i at each period p are independent, then we denote by $\overline{Y_i}$ and σ_i^2 the mean and the variance of the set M_i generated by the member node i, and by $\overline{Y}$ the mean of the pair of sets (M_i, M_j) generated by the member node i and j respectively as follows:

$$\overline{Y_i} = \frac{1}{Card_w(M_i)} \sum_{k=1}^{|M_i|} (y_{ik} \times wgt(y_{ik})), \sigma_i^2 = \frac{1}{Card_w(M_i)} \sum_{k=1}^{|M_i|} (wgt(y_{ik}) \times (y_{ik} - \overline{Y_i})^2),$$

$$\overline{Y} = \frac{1}{Card_w(M_i)} \sum_{k=1}^{|M_i|} (y_{ik} \times wgt(y_{ik})) + \frac{1}{Card_w(M_j)} \sum_{k=1}^{|M_j|} (y_{jk} \times wgt(y_{jk})),$$

where $y_{ik} \in M_i$ and $y_{jk} \in M_j$.

Since $Card_w(M_i) = Card_w(M_j) = T$:

$$\overline{Y_i} = \frac{1}{T} \sum_{k=1}^{|M_i|} (y_{ik} \times wgt(y_{ik})), \sigma_i^2 = \frac{1}{T} \sum_{k=1}^{|M_i|} (wgt(y_{ik}) \times (y_{ik} - \overline{Y_i})^2),$$

$$\overline{Y} = \frac{1}{2 \times T} \Big(\sum_{k=1}^{|M_i|} (y_{ik} \times wgt(y_{ik})) + \sum_{k=1}^{|M_j|} (y_{jk} \times wgt(y_{jk})) \Big),$$

where $y_{ik} \in M_i$ and $y_{jk} \in M_j$.

The total variation (ST), in a pair of sets, is the sum of the variation (SR) within each set and the variation (SF) between the sets. SF represents what is often called "explained variance" or "systematic variance". We can think of this as the variance that is due to the independent variable, the difference among the two sets. For example the difference between measures in two or more different sets. SR represents what is often called "error variance". This is the variance within sets, variance that is not due to the independent variable. For example, the difference between measures in the same set. The whole idea behind the analysis of variance, in a pair of sets, is to compare the ratio of the variance between the sets to the variance within each set in this pair. If the variance caused by the interaction between the measures, in a pair of sets, is much larger than the variance that appears within the sets, then it is because the means arent the same. Let us consider:

$$ST = SR + SF \Rightarrow$$

$$\sum_{l}^{\{i,j\}} \sum_{k=1}^{|M_l|} \left(wgt(y_{lk}) \times (y_{lk} - \overline{Y})^2\right) = \sum_{l}^{\{i,j\}} \sum_{k=1}^{|M_l|} \left(wgt(y_{lk}) \times (y_{lk} - \overline{Y_l})\right)^2 + \mathcal{T} \sum_{l}^{\{i,j\}} (\overline{Y_l} - \overline{Y})^2$$

(1)

4.2   Mean's Period Verification

In this section, we use three tests in the Anova model (Fisher, Tukey and Bartlett), to compute the means and the variances for every pair of sets, then to decide if the sets in this pair are redundant or not.

Fisher Test. The Fisher's test or F-test is a statistical hypothesis test for testing the equality of two variances by taking the ratio of the two variances and ensuring that this ratio does not exceed a certain theoretical value (find in Fisher's table). In the case of PSN, we compare, in a pair of sets, the ratio of the variance between the sets (SF) to that within each set in this pair (SR).

The general formula for the F-test is:

$$F = \frac{SF/(J-1)}{SR/(N-J)}$$

where J is the number of compared sets and N is the number of total measures in the compared sets. Therefore, J is equal to 2 in our case while N is equal to $2 \times \mathcal{T}$ (because $Card_w(M_i) = Card_w(M_j) = \mathcal{T}$).

Then, we deduce:

$$F = 2 \times (\mathcal{T} - 1) \times \frac{SF}{SR}$$

(2)

For each pair of sets, the CH will test the hypothesis that means of sets are the same or not. If the hypothesis is correct then, F will have a Fisher distribution, with $F(1, 2 \times (T - 1))$ degrees of freedom. The hypothesis is rejected if the F calculated from the measures is greater than the critical value of the F distribution for some desired false-rejection probability (risk α). Let $F_t = F_{1-\alpha}(1, 2 \times (T - 1))$.

The decision is based on F and F_t:

- if $F > F_t$ the hypothesis is rejected with false-rejection probability α, and the variance between the sets are significative.
- if $F \leq F_t$ the hypothesis is accepted.

Tukey Test. The Tukey's post-hoc test [25] is a single-step multiple comparison procedure and statistical test. It can be used to calculate the difference between the means of two or multiple sets. Tukey's test works by defining a value known as Honest Significant Difference (HSD). HSD represents the minimum distance between the means of two sets to be considered statistically significant.

Tukey's test can be applied to a pair of sets (M_i, M_j) based on the following equations:

$$SS_{total} = \sum_{l}^{\{i,j\}} \sum_{k=1}^{|M_l|} \left(wgt(y_{lk}) \times y_{lk}^2 \right) - \frac{\left(\sum_{l}^{\{i,j\}} \sum_{k=1}^{|M_l|} \left(wgt(y_{lk}) \times y_{lk} \right) \right)^2}{2 \times T} \tag{3}$$

$$SS_{among} = \frac{\left(\sum_{k=1}^{|M_i|} \left(wgt(y_{ik}) \times y_{ik} \right) \right)^2 + \left(\sum_{k=1}^{|M_j|} \left(wgt(y_{jk}) \times y_{jk} \right) \right)^2}{T}$$
$$- \frac{\left(\sum_{l}^{\{i,j\}} \sum_{k=1}^{|M_l|} \left(wgt(y_{lk}) \times y_{lk} \right) \right)^2}{2 \times T} \tag{4}$$

$$SS_{within} = SS_{total} - SS_{among}; \quad df_{among} = 1; \quad df_{within} = 2 \times T - 2$$

$$MS_{among} = \frac{SS_{among}}{df_{among}}; \quad MS_{within} = \frac{SS_{within}}{df_{within}}; \quad F = \frac{MS_{among}}{MS_{within}}$$

where:

- SS_{within} : Sum of squares within the pair of sets (M_i, M_j),
- SS_{among} : Sum of squares between the sets in the pair (M_i, M_j),
- MS_{within} : Mean squares within the pair of sets (M_i, M_j),
- SS_{among} : Sum of squares between the sets in the pair (M_i, M_j).

Therefore, when we calculate F we check to see if it is statistically significant based on studentized range distribution table with appropriate degrees of freedom $F_t = df(df_{among}, df_{within})$. The decision is based on F and F_t:

- if $F > F_t$ the hypothesis is rejected with false-rejection probability α, and the variance between the sets M_i and M_j are significative.
- if $F \leq F_t$ the hypothesis is accepted.

Bartlett Test. The Bartlett's test [26] is used to test if two or multiple data sets are from populations with equal variances. Equal variances across data sets is called homogeneity of variances. Some statistical tests, for example the analysis of variance, assume that variances are equal across data sets. The Bartlett test can be used to verify that assumption. Bartlett's test is used to test the null hypothesis, H_0 that variances of all data sets are equal against the alternative that at least two are different. In our case, we test the hypothesis H_0 for every pair of sets (M_i, M_j) each having a size T and with variances σ_i^2 and σ_j^2 respectively. Bartlett's test statistic is:

$$F = \frac{2 \times (T-1) \ln(\sigma_p^2) - (T-1)(\ln \sigma_i^2 + \ln \sigma_j^2)}{\lambda} \tag{5}$$

where :

$$\lambda = 1 + \frac{1}{2 \times (T-1)} \tag{6}$$

and σ_p^2 is the pooled variance, which is a weighted average of the period variances and it is defined as:

$$\sigma_p^2 = \frac{1}{2 \times (T-1)} \times (\sigma_i^2 + \sigma_j^2)$$

Bartlett's test has approximately a $(J-1)$ degrees of freedom where J is equal to 2 in our case. Thus the null hypothesis is rejected if $F > T_{J-1,\alpha}$ (where $T_{J-1,\alpha}$ is the upper tail critical value for the T_{J-1} distribution). We suppose that $F_t = T_{J-1,\alpha}$, thus the decision is based on the following rule:

- if $T > F_t$ the hypothesis is rejected with false-rejection probability α, and the variance between the sets M_i and M_j are significative.
- if $T \leq F_t$ the hypothesis is accepted.

4.3   Aggregation at the CH Level

In this section, we present the algorithms that follow each CH to find redundant data sets based on Anova model, then to remove redundancy before sending them to the sink.

Sets Redundancy Searching. In our technique, one way Anova model is used to find all pairs of sets that have low variance between their measures. Algorithm 1 describes how these pairs are found in our technique. For every pair of sets (M_i, M_j), we calculate the corresponding F score as described in each test presented before (line 4). Then, we search the corresponding threshold F_t based on the probability table for each test with the appropriate degrees of freedom (line 5). Finally, we conclude that M_i and M_j are redundant sets in the case where the variance between their measures (F) is less than the threshold F_t (line 6).

Algorithm 1. CH aggregation algorithm.

Require: Set of measures' sets $M = \{M_1, M_2...M_n\}$.
Ensure: All pairs of sets (M_i, M_j), such that $F \leq F_t$.
1: $S \leftarrow \emptyset$
2: **for** each set $M_i \in M$ **do**
3:     **for** each set $M_j \in M$ such that $M_j \neq M_i$ **do**
4:         compute F for (M_i, M_j)
5:         find F_t
6:         **if** $F \leq F_t$ **then**
7:             $S \leftarrow S \cup \{(M_i, M_j)\}$
8:         **end if**
9:     **end for**
10: **end for**
11: **return** S

Algorithm 2. Selecting sets algorithm.

Require: All pairs of sets (M_i, M_j), such that $F \leq F_t$.
Ensure: List of selected sets, L.
1: $L \leftarrow \emptyset$
2: **for** each pair of sets (M_i, M_j) **do**
3:     Consider $|M_i| \geq |M_j|$
4:     $M_i \leftarrow \text{sort}(M_i, |M_i|)$, M_i is sorted in increasing order of the measures
5:     **for** $k = 1 \rightarrow |M_j|$ **do**
6:         Search similar of $M_j[k]$ in M_i
7:         find $M_i[l] / Similar(M_j[k], M_i[l]) = 1$
8:         **if** $M_i[l]$ exists **then**
9:             $wgt(M_i[l]) \leftarrow wgt(M_i[l]) + wgt(M_j[k])$
10:        **else**
11:            $M_i \leftarrow M_i \cup \{(M_j[k], wgt(M_j[k]))\}$
12:        **end if**
13:    **end for**
14:    $L \leftarrow L \cup \{M_i\}$
15:    Remove all pairs of sets containing one of the two sets M_i and M_j
16: **end for**

Redundant Sets Reduction. After identifying all pairs of redundant sets, the CH deletes redundant data sets sent from neighboring sensors in order to reduce the amount of data transmitted to the sink while conserving the integrity of information. Algorithm 2 shows how the CH selects the data sets to be sent to the sink among the pairs of redundant received sets. For each similar pair of set, the CH chooses the one having the highest cardinality (line 3), then it sorts it in increasing order of the measures to accelerate a measure search[1]. After that,

---

[1] In our experiments we used the binary search.

for each measure in the other set, CH searches for its similar in the highest set and merges its weight to the similar one found (line 9). Otherwise, CH adds the measure with its weight to the highest set (line 11). The objective of merging the weights of similar measures is to save the information without any loss. Finally, the CH removes all pairs of redundant sets that contain M_i or M_j from the set of pairs (which means it will not check them again) (line 15).

5   Performance Evaluation

In this section, we present the experimental results which evaluate the performance of our proposed technique. The objective of these experiments is to confirm that our technique can successfully achieve desirable results for energy conservation in PSNs. Therefore, we used the publicly available Intel Lab dataset which contains data collected from 46 sensors deployed in the Intel Berkeley Research Lab [27]. Mica2Dot sensors with weather boards collect timestamped topology information, along with humidity, temperature, light and voltage values once every 31 s. The data was collected using TinyDB in-network query processing system built on the TinyOS platform. In our experiments, we used a file that includes a log of about 2.3 million readings collected from these sensors. For the sake of simplicity, in this paper we are interested in one field of sensor measurements: the temperature. We assume that all nodes send their data to a common CH placed at the center of the Lab. First, each node reads periodically real measures while applying the member node aggregation. At the end of this step, each node sends its set of measures/weights to the CH which in turn applies CH aggregation to theses sets. Furthermore, we compare our technique to the PFF technique proposed in [21] with two values of the Jaccard similarity threshold t (0.75 and 0.8). We have implemented both techniques on a java simulator and we compared the results of 15 periods in all the experiments.

We evaluated the performance using the following parameters:

- δ, which defines the *Similar* function between two measurements. We varied δ to : 0.03, 0.05, 0.07 and 0.1.
- T, the number of sensor measurements taken by each sensor node during a period. We varied T to: 200, 500, 1000 and 2000.
- α, the false-rejection probability in the Anova model which we varied to 0.01 and 0.05.

5.1   Percentage of Data Sent to the CH

In the first aggregation level, each member node searches the similarity between measures captured at each period, using the *Similar* function, and assigns for each measure its weight. Therefore, the result of the aggregation in this level depends on the chosen threshold δ, and the number of the collected measures in period T. Figure 3 shows the percentage of data sent by each node to the CH at each period with and without applying the first aggregation level.

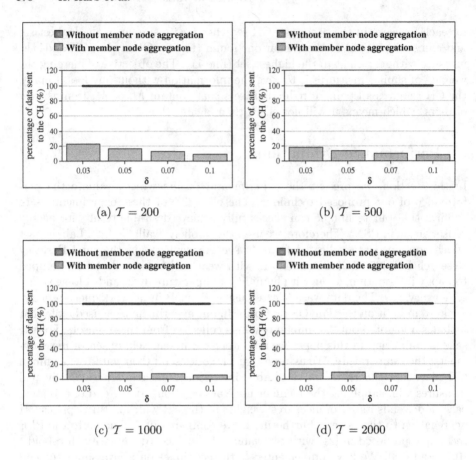

Fig. 3. Percentage of data sent to the CH.

The obtained results show that, at each period, each node reduces more than 68 % the amount of collected data after the first aggregation level while it sends all the collected data, e.g. 100 %, without applying this aggregation level. Therefore, our technique can successfully eliminate redundant measures at each period and reduce the amount of data sent to the CH. We can observe also that at the first aggregation level, data redundancy increases when $\mathcal{T}$ or δ increases. This is because, *Similar* function will find more similar measures to be eliminated at each period.

5.2   Number of Pairs of Redundant Sets Generated at the CH

When receiving all the sets from its member nodes at the end of each period, CH applies the second aggregation level to find pairs of redundant sets. Figure 4 shows the obtained number of pairs of redundant sets when applying one way Anova model with the three tests presented above, compared to the number of

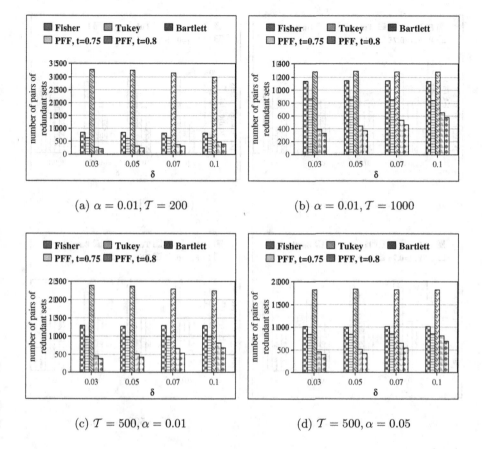

Fig. 4. Number of pairs of redundant sets.

similar sets obtained when applying PFF. In Fig. 4(a and b), we fixed α to 0.01 and we varied T to 200 and 1000 respectively, while in Fig. 4(c and d) we fixed T to 500 and we varied α to 0.01 and 0.05 respectively. The obtained results show that, CH finds more redundant sets when applying our technique in all the cases. This is because, the variance condition in the one way Anova model is more flexible compared to the similarity condition used in PFF.

Based on the obtained results, we can also deduce:

- Bartlett test finds more pairs of redundant sets compared to Tukey and Fisher tests. This is because Bartlett test is more flexible regarding the variance between measures (Eq. (5)) compared to the variance calculated in Fisher (Eq. (2)) and Tukey (Eqs. (3) and (4)).
- The obtained number of pairs of redundant sets decreases in the three tests when α increases. This is because, when the risk α increases the null hypothesis will have higher probability of being rejected.

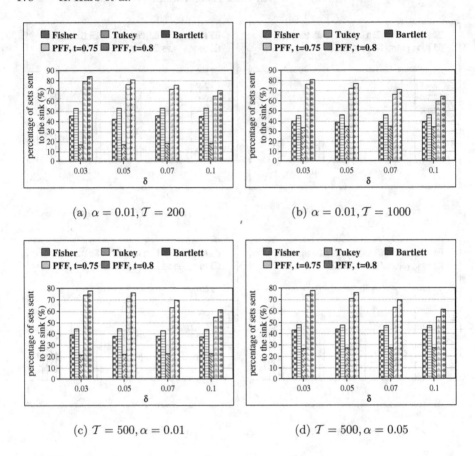

Fig. 5. Percentage of sets sent to the sink.

5.3   Percentage of Sets Sent to the Sink

In this section, our objective is to show how the CH is able to eliminate redundant
sets at each period using redundant data reduction algorithm. Figure 5 shows
the percentage of the remained sets that will be sent to the sink after eliminating
the redundancy. Figure 5(a and b) show the results when we fixed α to 0.01 and
varied $\mathcal{T}$ to 200 and 1000 respectively, while Fig. 5(c and d) show the results
when we varied α to 0.01 and 0.05 and fixed $\mathcal{T}$ to 500. We can show clearly that,
our technique sends much less sets at each period to the sink with the different
parameters. This is because, CH found more redundant sets using the variance
condition (Fig. 4).

Based on the obtained results, we can also deduce:

- Bartlett test sends the less percentage of sets to the sink since it found more
  redundant sets compared to Fisher and Tukey tests (see Fig. 4).

- The percentage of sets sent to the sink for the three tests is almost fix when fixing $\mathcal{T}$ and increasing δ. This is because, the data set saves the same variance when changing δ.
- CH eliminates more redundant sets in the three tests when decreasing α. This is because when α decreases, the number of pairs of redundant sets increases (see Fig. 4(c and d)).

5.4   Data Accuracy

Eliminating redundant data without losing accuracy is an important challenge for the WSN. Data accuracy represents the measure "loss rate" taken by sensor nodes and not received by the sink [21]. Since CH merges the weights of similar measures in the redundant sets in one record compared to PFF which removes one between them, the integrity of the information is totally saved in our technique. This fact is obtained independently from the values of $\mathcal{T}$, δ and α, whereas the percentage of loss measures in PFF can up to 5.4 for some values of the parameters [21]. Therefore, we can consider that our technique decreases the amount of redundant data forwarded to the sink without any loss of information integrity.

5.5   Energy Consumption at the CH

In this section, our objective is to study the energy cost at the CH level. Therefore, we used the same radio model as discussed in [27]. In this model, a radio dissipates $E_{elec} = 50\,nJ/bit$ to run the transmitter or receiver circuitry and $\beta_{amp} = 100\,pJ/bit/m^2$ for the transmitter amplifier. Radios have power control and can expend the minimum required energy to reach the intended recipients as well as they can be turned off to avoid receiving unintended transmissions. Equations used to calculate transmission costs and receiving costs for a k-bit messages and a distance d are respectively shown in Eqs. (7) and (8):

$$E_{TX}(k, d) = E_{elec} \times k + \beta_{amp} \times k \times d^2 \tag{7}$$

$$E_{RX}(k) = E_{elec} \times k \tag{8}$$

Recall that the CH will receive n data sets coming from its member nodes at each period. The size of each set is equal to the number of measures sent in addition to the number of weights sent. We consider that each measure or weight is equal to 64 bits. Therefore, the energy consumption at the second level will be equal to the energy consumed when the CH receives the data sets from its member in addition to the energy consumed when it sends them after the aggregation. Consequently, after 15 periods as we calculated in our experiments, the total energy consumption at the CH is calculated as shown in Eq. (9)

$$E_{CH}(m, d) = E_{RX_{total}} + E_{TX}(m, d) = \left(2 \times 64 \times E_{elec} \times \sum_{i=1}^{n} |M_i|\right) +$$

$$\left(64 \times E_{elec} \times m + 64 \times \beta_{amp} \times m \times d^2\right) \tag{9}$$

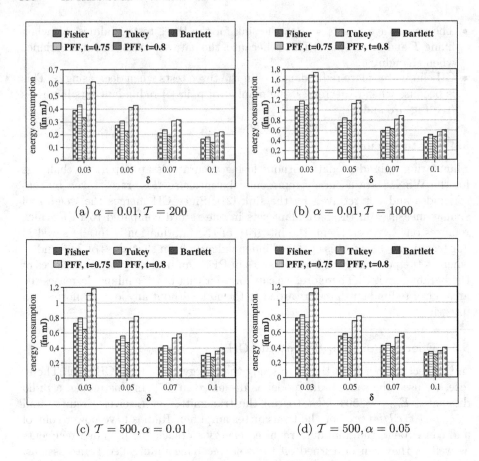

(a) $\alpha = 0.01, \mathcal{T} = 200$ (b) $\alpha = 0.01, \mathcal{T} = 1000$

(c) $\mathcal{T} = 500, \alpha = 0.01$ (d) $\mathcal{T} = 500, \alpha = 0.05$

Fig. 6. Energy consumption at the CH.

where m is the total number of the measures with their weights after the aggregation in all the sets and d is the distance between the CH and the sink.

Figure 6 shows the energy consumption comparison between our technique and the PFF at the CH level when fixing α and varying $\mathcal{T}$ (Figs. 6(a and b)) and when fixing $\mathcal{T}$ and varying α (Figs. 6(c and d)). The obtained results show that our technique minimizes the energy consumption of the CH up to 45 % when compared to the PFF. These results are obtained due to the fact that our technique eliminates more redundant sets compared to PFF (see Fig. 5). Therefore, we can consider that our technique decreases the amount of redundant data forwarded to the sink and performs an overall lossless process in terms of information and integrity by conserving the weight of each measure.

Based on the obtained results, we can also deduce:

- Bartlett test decreases energy consumption of the CH more than the other tests.

- The energy consumption at the CH is more minimized when α decreases. This is because, when α decreases the percentage of sets sent to the sink decreases (see Fig. 5(c and d)).

5.6   Discussion

In this section, we discuss the results for the three tests used with ANOVA model in terms of conserving energy of the sensors. First, by fixing α and varying T as shown in Fig. 6(a, b and c), we can deduce that Bartlett test allows more energy saving than Fisher and Tukey tests when the period is small (e.g. T equals to 200 and 500 in Fig. 6(a and c). Contrarily, Fisher test gives better results for large periods, e.g. T is greater than 1000 in Fig. 6c. This is because, Bartlett test is more flexible regarding the variance between measures in small periods (Eq. (5)) while Fisher test is more flexible in large periods (Eq. (2)).

On the other hand, by fixing T and varying α as shown in Fig. 6(c and d), the energy consumption is more minimized in the three tests when α is small, e.g. $\alpha = 0.01$ in Fig. 6c. This is because, the energy consumption highly depends on the number of pairs of redundant sets eliminated which increases when α decreases. Consequently, the null hypothesis will have higher probability of being rejected when α decreases. Furthermore, the general trend observed that is Bartlett test gives better results, in terms of energy consumption, when T is small while Fisher test gives better results when T is large.

6   Conclusion and Future Work

In this paper, we proposed a new technique for data aggregation in PSN that enforces both energy consumption and integrity of the aggregated data. Our proposed technique consists of two-level of data aggregation which applies at each cluster in a clustering network architecture. The first level is applied at the node itself to eliminate redundancy from the collected raw data before sending them to the CH. At the second level, CH searches nodes that generate redundant data sets based on the dependence of conditional variance with three different Anova tests. Comparing to other existing data aggregation techniques, experimental results on real sensor data show the effectiveness of our technique in terms of energy consumption and information integrity.

A direction for future work is to adapt our proposed technique to take into consideration reactive periodic sensor networks, where sensor nodes operate with different sampling rate. In periodic applications the dynamics of the monitored condition or process can slow down or speed up; and to save more energy the sensor node can adapt its sampling rates to the changing dynamics of the condition or process.

References

1. Abbasi, A., Younis, M.: A survey on clustering algorithms for wireless sensor networks. J. Comput. Commun. **30**(14–15), 2826–2841 (2007)
2. Rozyyev, A., Hasbullah, H., Subhan, F.: Indoor child tracking in wireless sensor network using fuzzy logic technique. Res. J. Inf. Technol. **3**(2), 81–92 (2011)
3. Sabri, N., Aljunid, S.A., Ahmad, R.B., Yahya, A., Kamaruddin, R., Salim, M.S.: Wireless sensor actor network based on fuzzy inference system for greenhouse climate control. J. Appl. Sci. **11**(17), 3104–3116 (2011)
4. Aslan, Y.E., Korpeoglu, I., Ulusoy, O.: A framework for use of wireless sensor networks in forest fire detection and monitoring. Comput. Environ. Urban Syst. **36**(6), 614–625 (2012)
5. Qian, H., Sun, P., Rong, Y.: Design Proposal of Self-Powered WSN Node for Battle Field Surveillance,. In: Energy Proced., Part B, vol. 16, pp. 753–757 (2012)
6. Padmavathi, G., Shanmugapriya, D., Kalaivani, M.: A study on vehicle detection and tracking using wireless sensor networks. Wirel. Sens. Netw. **2**(2), 173–185 (2010)
7. Di Pietro, R., Michiardi, P., Molva, R.: Condentiality and integrity for data aggregation in WSN using peer monitoring. Secur. Comm. Netw. **2**(2), 181–194 (2009)
8. Azhar, M., Ke, S., Shaheen, K., Mi, X.: Data mining techniques for wireless sensor networks: a survey. Int. J. Distrib. Sens. Netw. 2013 (2013), 24 pages (2013)
9. Mirhadi, P., Zandinia, S., Goodarzipour, A., Salimi, S., Goodarzipour, H.: IP2P K-means: an efficient method for data clustering on sensor networks. Manage. Sci. Lett. **3**(3), 967–972 (2013)
10. Yuan, F., Zhan, Y., Wang, Y.: Data density correlation degree clustering method for data aggregation in WSN. IEEE Sensors J. **14**(4), 1089–1098 (2014)
11. Enam, R.N., Qureshi, R., Misbahuddin, S.: A uniform clustering mechanism for wireless sensor networks. Int. J. Distrib. Sensor Netw. 2014 (2014), 14 pages (2014)
12. Tripathi, A., Gupta, S., Chourasiya, B.: Survey on data aggregation techniques for wireless sensor networks. Int. J. Adv. Res. Comput. Commun. Eng. **3**(7), 7366–7371 (2014)
13. Nokhanji, N., Hanapi, Z.M.: A survey on cluster-based routing protocols in wireless sensor networks. J. Appl. Sci. **14**(18), 2011–2022 (2014)
14. Zou, P., Liu, Y.: A data-aggregation scheme for WSN based on optimal weight allocation. J. Netw. **9**(1), 100–107 (2014)
15. Tran, K.T.-M., Oh, S.-H.: A data aggregation based efficient clustering scheme in underwater wireless sensor networks. In: Jeong, Y.-S., Park, Y.-H., Hsu, C.-H.R., Park, J.J.J.H. (eds.) Ubiquitous Information Technologies and Applications. LNEE, vol. 280, pp. 541–548. Springer, Heidelberg (2014)
16. Kumar, S., Prateek, M., Ahuja, N.J., Bhushan, B.: MEECDA: multihop energy efficient clustering and data aggregation protocol for HWSN. Int. J. Comput. Appl. **88**(9), 28–35 (2014)
17. Chao, C.-M., Hsiao, T.-Y.: Design of structure-free and energy-balanced data aggregation in wireless sensor networks. J. Netw. Comput. Appl. **37**, 229–239 (2014)
18. Li, G., Wang, Y.: Automatic ARIMA modeling-based data aggregation scheme in wireless sensor networks. EURASIP J. Wirel. Commun. Netw. **2013**(1), 1–13 (2013)
19. Shan, M., Chen, G., Luo, D., Zhu, X., Wu, X.: Building maximum lifetime shortest path data aggregation trees in wireless sensor networks. J. ACM Trans. Sensor Netw. (TOSN) **11**(1), Article 11 (2014)

20. Shim, Y., Kim, Y.: Data Aggregation with multiple sinks in Information-Centric Wireless Sensor Network. In: International Conference on Information Networking (ICOIN 2014), pp. 13–17 (2014)
21. Bahi, J., Makhoul, A., Medlej, M.: A two tiers data aggregation scheme for periodic sensor networks. Ad Hoc Sensor Wirel. Netw. **21**(1–2), 77–100 (2014)
22. Harb, H., Makhoul, A., Tawil, R., Jaber, A.: A suffix-based enhanced technique for data aggregation in periodic sensor networks. In: 10th IEEE Internationl Conference on Wireless Communications and Mobile Computing (IWCMC 2014), pp. 494–499, Cyprus (2014)
23. Harb, H., Makhoul, A., Laiymani, D., Jaber, A., Tawil, R.: K-Means based clustering approach for data aggregation in periodic sensor networks. In: 10th IEEE International Conference on Wireless and Mobile Computing, Networking and Communications (WIMOB 2014), pp. 434–441, Cyprus (2014)
24. Laiymani, D., Makhoul, A.: Adaptive data collection approach for periodic sensor networks, In: 9th IEEE International Conference on Wireless Communications and Mobile Computing (IWCMC), pp. 1448–1453, Italy (2013)
25. Hall, R.: Psychology World (1998). http://web.mst.edu/~psyworld/tukeysexample.htm
26. Snedecor, G., Cochran, G.: Statistical Methods, 8th edn. Iowa State University Press, Ames (1989)
27. Samuel Madden. http://db.csail.mit.edu/labdata/labdata.html

Author Index